**TRICKS OF
THE TRADE**

Plumbing

Ernest Hall, FRSH

Pelham Books

First published in Great Britain by
PELHAM BOOKS LTD
44 Bedford Square
London WC1
1979

ISBN 0 7207 1170 3

Printed and bound in Hong Kong.

Plumbing

Contents

List of Line Illustrations   6

Introduction   9

Part One: Basic Maintenance and Repair
1   Taps and Stop-cocks   11
2   Ball-valves   22
3   The Lavatory Suite   30
4   Hot-water Supply   38
5   Corrosion and Frost   47
6   The Drains   60

Part Two: Plumbing Techniques
7   Handling Copper, Stainless-steel and PVC
     Tubing   68

Part Three: Some Plumbing Projects
8   Fitting an Outside Tap   76
9   Plumbing in an Automatic Washing-
     machine   81
10   Renewing Sinks, Baths and Washbasins   86
11   Installing a Shower   96
12   The Lavatory Suite: Conversion to Low
     Level, or Renewal of a Pan   102

Index   109

Line
Illustrations

		page
Fig. 1	The basic bib-tap	11
Fig. 2	The main stop-cock may be under the kitchen sink or in an external purpose-made pit	12
Fig. 3	A pillar-tap with easy-clean cover	14
Fig. 4	Modern pillar-tap with shrouded head	16
Fig. 5	Cross-section of the nozzle of a supatap	17
Fig. 6	Nylon full-stop washer and jumper set	18
Fig. 7	A Portsmouth-pattern ball-valve	23
Fig. 8	Equilibrium ball-valve	27
Fig. 9	BRS or diaphragm ball-valve with overhead outlet	28
Fig. 10	Modern direct-action flushing cistern	31
Fig. 11	An old-fashioned 'Burlington' flushing cistern	32
Fig. 12	A double-trap siphonic lavatory suite	36
Fig. 13	Simple cylinder-storage hot-water system	39
Fig. 14	Indirect cylinder storage hot-water system	44
Fig. 15	Sacrificial anode installed in cold-water storage cistern	50
Fig. 16	Installing a sacrificial anode in a hot-water storage tank	51
Fig. 17	The feed and expansion tank of an indirect hot-water system	54
Fig. 18	Two ways of protecting the overflow pipe of a cistern from frost	56
Fig. 19	Making a temporary repair to a burst lead pipe	59

Fig. 20 Clearing a blocked sink waste with a force
         cup   61
Fig. 21 Means of access to U-traps and bottle-traps   62
Fig. 22 Clearing an intercepting trap by plunging   66
Fig. 23 Making a Type 'A' non-manipulative
         compression joint   69
Fig. 24 The use of a tube cutter with reamer will
         ensure a clean, square-ended pipe end   70
Fig. 25 Making an integral-ring soldered capillary
         joint   71
Fig. 26 Fitting a compression 'tee' into a rising
         main to provide a supply for an outside tap   78
Fig. 27 Providing the water supply to an outside
         tap   79
Fig. 28 The outside tap fitted   80
Fig. 29 Conventional washing-machine installation
         with stand-pipe outlet   82
Fig. 30 Fitting a 'Kontite' hose-connector   84
Fig. 31 The plastic 'Opella' hose-connector   85
Fig. 32 A 'top hat' or 'spacer' washer must be used
         when fitting a tap to a modern stainless-
         steel or enamelled-steel sink   88
Fig. 33 When installing a ceramic washbasin, the
         slot in the waste fitting must coincide with
         the built-in overflow of the basin   91
Fig. 34 A combined bath waste and overflow fitting   94
Fig. 35 'Copperbend' – either with or without a tap
         connector – can be useful for making
         'out-of-sight' bends in copper tubing   95
Fig. 36 Design requirements for a conventional
         shower installation with hot-water supply
         derived from a cylinder storage hot-water
         system   97
Fig. 37 The 'Flomatic' electric pump unit can be
         used to boost pressure to a shower where
         the cold-water storage cistern is too
         low to permit gravity operation   100

7

Fig. 38 The pan of a low-level lavatory suite
        must be sufficiently far from the wall
        behind it to permit the flushing cistern
        to be accommodated   102
Fig. 39 'Multikwik' plastic drain-connectors
        provide a simple means of connecting a
        lavatory pan to a branch drain or soil pipe 105
Fig. 40 To remove and replace the pan of a ground-
        floor lavatory suite it will usually be
        necessary to break the pan outlet just
        behind the trap   106

Introduction

A great many householders who claim, with justification, to be 'handy about the house' phone for a professional plumber at the least hint of trouble with their domestic hot- and cold-water or drainage systems. They are afraid of beginning a job that they are unable to finish – with disastrous results!

Yet modern materials and methods – and a knowledge of a few 'tricks of the trade' – make it easy for the do-it-yourself enthusiast to carry out his own plumbing repairs and maintenance, and to undertake such plumbing projects as converting an old-fashioned high-level lavatory suite to low-level operation, replacing sinks, baths and washbasins, installing a shower, providing a garden water supply and plumbing in an automatic washing-machine or dishwasher.

In this book – starting off with basic plumbing jobs like servicing taps, stop-cocks and ball-valves, and protecting the plumbing system from frost and corrosion – I give step-by-step instructions, easily followed by an in-experienced householder with a minimal toolkit, for carrying out plumbing work about the house *in a professional manner*.

Tricks of the trade are easy to spot in the text as they are set out in bold type and marked with the symbol ➤.

Part One

Basic Maintenance and Repair

1 Taps and Stop-cocks

The tap is an item of domestic plumbing equipment in constant daily use. Consequently it is the piece of equipment that is most likely to need occasional maintenance if it is to continue to function efficiently.

Illustrated is a cross-section of the most basic kind of tap – the brass bib-tap. You may have bib-taps like this projecting from the glazed tiles above your kitchen sink. In a modern home, however, it is probable that an extremely simple tap of this kind will be found only against an external wall to provide a garden water supply.

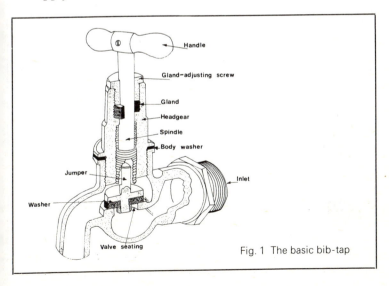

Fig. 1 The basic bib-tap

However, with the exception of the supatap (which will be dealt with later), all modern taps – however different they may be in appearance – operate on the same principle as this straightforward appliance.

As can be seen, turning the tap handle raises or lowers a washered valve or 'jumper'. When the tap is turned off, this valve is pressed firmly against the valve seating to prevent the flow of water.

The tap spindle, which connects the handle to the valve, passes through a 'gland' or 'stuffing box'. This is packed with greased wool to prevent water from flowing up the spindle when the tap is opened. In some modern taps a rubber 'O' ring-seal replaces this traditional gland.

Re-washering

The need to renew a tap washer is indicated by a constant drip after the tap has been turned off. The onset of this

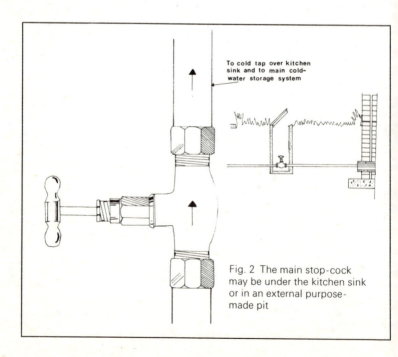

To cold tap over kitchen sink and to main cold-water storage system

Fig. 2 The main stop-cock may be under the kitchen sink or in an external purpose-made pit

drip is insidious and the instinctive reaction of the household is to attempt to turn the tap off harder. This may work for a while, but the trouble will recur and further attempts to force the tap to close could damage it.

The first step to take in rewashering a tap is to turn off the water supply to it.

The cold tap over the kitchen sink, and any garden tap, will be connected directly to the water main. To cut off the water supply you must turn off the main stop-cock. You will probably find this on the rising main under the kitchen sink. In some older houses, however, the only stop-cock may be in a purpose-made pit, with a hinged metal cover, in the front garden or, just possibly, in the pavement outside the house.

Bathroom cold taps are also sometimes connected directly to the rising main. But in most cases the bathroom cold taps will be supplied from the cold-water storage cistern, probably situated in the roof space. This cistern will also supply the hot-water system and thus, ultimately, the hot-water taps.

To cut off the water supply from any tap supplied ← **from a main storage cistern, there is no need to turn off the main stop-cock. The cold supply to the kitchen sink need not be interrupted by your washer-changing operation.**

Place a slat of wood across the top of the storage cistern and tie the float arm of the ball-valve to it so as to prevent water flowing into the cistern through the ball-valve. You can now open the taps to drain the cistern. When water ceases to flow, the washer can be changed.

One further tip: when changing the washer on a hot-water tap there is no need to drain all the hot water from the hot-water storage cylinder, provided that the bathroom cold taps are supplied from the storage cistern.

After tying up the cistern ball-valve, open up the

bathroom cold taps only. Open up the hot taps only when water ceases to flow from the cold ones. A pint or two of water will drain off and then flow will cease. The storage cylinder will remain full of hot water. This is because the supply pipes to the hot taps are taken from the vent pipe ABOVE the hot-water cylinder.

The next step in rewashering is to remove the headgear of the tap. You'll need a reliable wrench for this. **The jaws of the wrench must fit tightly and securely to the 'flats' of the headgear, and the longer the handle the more leverage you'll be able to obtain.**

Modern taps usually have an easy-clean cover. With the tap fully opened it is possible to raise this cover high enough to slip the jaws of the wrench under it to grip the flats of the tap.

You should be able to unscrew and raise the cover by hand. If you are compelled to use a wrench on it, pad the jaws well with cloth to avoid damaging the chromium finish.

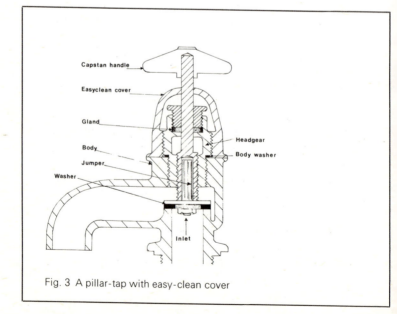

Fig. 3 A pillar-tap with easy-clean cover

Having unscrewed and removed the tap's headgear, you may now find the jumper resting on the valve seating. The washer is secured to the jumper by means of a tiny nut. This is sometimes difficult to undo. **Apply a** ◄─── **drop of penetrating oil and then, after a brief interval, secure the jumper firmly in a vice and tackle the nut with a spanner of the correct size. If the nut still seems to be immoveable you can buy a new jumper and washer complete from any d-i-y store.**

Make sure that your new washer – or washer and jumper set – is the right size. If you tell your supplier which tap you require it for – sink, basin or bath as the case may be – he'll sell you the right one.

When renewing the washer of a bathroom tap, or of the hot tap over the kitchen sink, you may find that the jumper is not resting on the valve seating when the headgear is unscrewed. It is 'pegged' into the headgear. It can be turned round and round but cannot be removed.

In this event you'll have to make a really determined effort to unscrew the retaining nut to enable the washer to be changed. However, even if it seems firmly welded in position there's still a 'trick of the trade' that will solve the problem.

Insert the blade of a screwdriver between the plate ◄─── **of the jumper and the headgear and forcibly break the pegging. Now you can remove the jumper. Buy a new jumper and washer complete, but, before inserting it in the headgear of the tap, 'burr' the stem with a rasp or coarse file so that it gives an 'interference fit'.**

Perhaps your tap has a 'shrouded head' (see overleaf), and the handle and headgear appear to form one unit. Once the 'shrouded head' has been removed, you'll find that the interior is just like any other tap.

The way in which the head is removed depends upon the make of the tap. Some shrouded heads can be simply pulled off. Some are secured by a tiny grub-screw in the

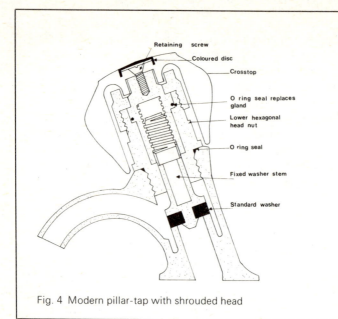

Retaining screw
Coloured disc
Crosstop
O ring seal replaces gland
Lower hexagonal head nut
O ring seal
Fixed washer stem
Standard washer

Fig. 4 Modern pillar-tap with shrouded head

side, similar to the grub-screw that retains the 'crutch' or 'capstan' head of a conventional tap. With yet another make the tap has to be fully opened and then given another final half-turn that allows the head to be pulled off.

Finally – and this is perhaps the commonest method by which shrouded heads are secured – there may be a retaining screw concealed under the plastic HOT or COLD indicator at the top of the head.

It is wise to try the other methods of removal before prizing off this plastic indicator.

The 'supatap' is the trade name of a tap of totally different design from those already described. Manufactured by Deltaflow Ltd of Crawley, one of its main attractions to the handyman is the ease with which washer-changing can be carried out without the need to cut off the water supply.

Supataps are turned on and off by turning the actual

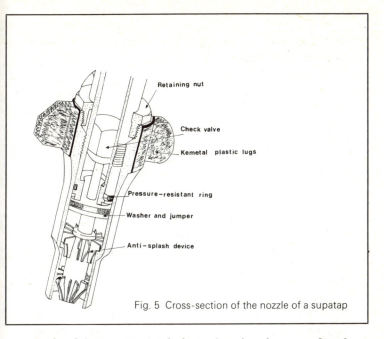

Retaining nut

Check valve

Kemetal plastic lugs

Pressure—resistant ring

Washer and jumper

Anti—splash device

Fig. 5 Cross-section of the nozzle of a supatap

nozzle of the tap. Kemetal plastic 'ears', or lugs, are fitted to the nozzle to make this a simple and comfortable operation.

To rewasher a supatap you must first of all unscrew ← the retaining nut at the top of the nozzle. Then, turn the tap on – and keep on turning. Water flow will increase at first but will then suddenly cease, just before the nozzle comes off in your hand, as a check-valve within the body of the tap falls into position.

Tap the nozzle outlet on a hard surface – not one that this operation would damage! Then turn it upside down and the anti-splash device, with the washer and jumper inside it, will fall out. Prize the washer and jumper out of the anti-splash device with a coin or a penknife blade and press in the replacement. Reassemble the tap.

That's all there is to it, but when you screw the nozzle back into position, remember that it has a left-hand thread. Overlook this and you'll have some anxious moments while you're wondering why 'the thread won't bite'.

Faulty Valve Seatings

It sometimes happens that a tap will begin to drip again immediately – or almost immediately – after it has been rewashered. This indicates that the valve seating has become worn and scored by grit from the water main and is no longer giving a watertight seal.

Plumbers have reseating tools which can be used to scrape the seating and thus to recondition it.

The handyman's best solution to this problem is to purchase and fit one of the nylon washer and seating sets that are available in handy packs from most d-i-y shops.

The tap is dismantled (after cutting off the water supply of course) and the new nylon seating placed squarely over the old metal one. The new plastic washer and jumper unit is placed in position in the headgear of the tap, which is then reassembled. All that remains to be done is to turn the tap off – hard!

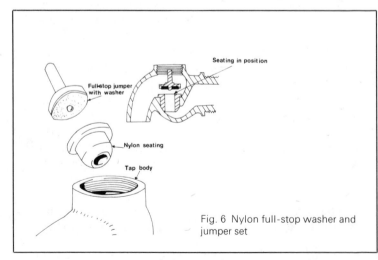

Fig. 6 Nylon full-stop washer and jumper set

This will force the new seating into position over the ←
old one. I have found that after fitting one of these
washer and seating sets the tap may continue to drip
slightly for a short while. However, after a week or
so's constant use, the new seating will settle into
position to give a watertight seal.

The valve seatings of supataps cannot, of course, be
reconditioned in this way. However, a simple tool for
renewing the valve seating of a supatap is readily
available from the manufacturers.

Troubles with the Gland

An escape of water up the spindle is a fault most likely to
be encountered in the taps over the kitchen sink. There
are two common causes.

Detergent-charged water from the hands of the
housewife using the taps may run down the spindle and
wash the grease out of the gland packing. Or the
connection of a hose – to water the garden or to fill a
washing-machine – to the tap nozzle can produce an
internal back pressure that will force water past the
gland.

The tap will turn off much too easily. It may even be
possible to 'spin' it on and off with a flick of the fingers.
Water hammer, a series of reverberating thuds following
the closing of a tap, can result from this. Water hammer is
caused by shock-waves created when the flow of water
through a pipe is suddenly stopped. Apart from the noise
nuisance involved, these shock-waves can damage
plumbing systems and produce leaks.

Faced with water escaping up the tap spindle, the
householder's first move should be to adjust the gland
packing nut. This is the first nut through which the
spindle of the tap passes. If the tap has an easy-clean
cover it will be necessary to remove the tap handle and
this cover to gain access to it.

Crutch and capstan handles are secured to the tap

spindle by means of a tiny grub-screw. **Remove this and put it in a safe place. The handle may now be readily removable – perhaps with the help of a few gentle taps from underneath.**

A useful tip for removing a stubborn handle is to unscrew the easy-clean cover and open the tap up fully. Raise the easy-clean cover and insert two pieces of wood between the base of the cover and the body of the tap. Now try to turn the tap off. The upward pressure of the easy-clean cover will force off the handle as you turn it downwards.

If you adopt this course of action, you should turn off the water supply to the tap first. You won't otherwise be able to stop the flow of water when you have removed the handle. If, however, you have managed to remove the tap handle by more conventional means without opening the tap, there will be no need to turn off the water supply when adjusting or renewing the gland.

To adjust the gland nut, give it half a turn in a clockwise direction. Turn the tap on and see if flow up the spindle has ceased. If it has not, tighten again.

Eventually, all the adjustment will be used up and the gland will need to be repacked.

Unscrew and remove the gland packing nut. Rake all existing packing material out of the gland with the point of a penknife. Repack with household wool steeped in Vaseline. Caulk down hard before replacing the gland nut and tightening it.

Very modern taps are likely to have 'O' ring-seals instead of a conventional gland. These are far less likely to leak, but if they do, replacement of the 'O' ring provides a simple cure.

Stop-cocks

Screw-down stop-cocks resemble taps in all respects and can be regarded as bib-taps set into a run of pipe. Since

20

they are rarely closed, their washers seldom need changing. This is fortunate since to cut off the water supply to the main stop-cock it would probably be necessary to seek the help of the local water authority.

Leakage past the gland is a more common fault and should receive immediate attention because leakage onto a wooden floor can, over a period of time, produce dry rot. Adjustment or renewal of the gland packing is carried out exactly as for taps. Turn the stop-cock off before attempting to renew the packing.

Jamming, as a result of long disuse, is a common fault in stop-cocks. You attempt to turn it off and find that it is apparently immovable. Applying penetrating oil and trying again, over a period of several days, may free it, but – in the kind of plumbing emergency that demands the closure of the stop-cock – you won't have several days to spare.

The best course of action is to make sure that the stop-cock doesn't jam. **Make a point of turning it on and off** ← **several times about twice a year. Then, when you open it for the last time, open it up fully and then give the handle just a quarter of a turn towards closure.** This won't materially affect the flow of the water but it will make it much less likely that the stop-cock will jam.

Mini-stop-cocks, operated by means of a screwdriver or the edge of a coin, are sometimes fitted into pipe runs just before they connect to taps or ball-valves. These permit water flow to be reduced if required. They also make it possible to cut off the water supply to a tap for washer-changing without the need to drain the whole system.

Gate-valves resemble screw-down stop-cocks in appearance and are used to control the flow of water at low pressure. The waterway is closed by a metal plate or 'gate'. When the valve is opened, this gate rises into the upper part of the valve. Gate-valves give a metal-to-metal seal that is less watertight than that afforded by the washer and jumper of a screw-down stop-cock. When open, however, they give an absolutely unimpeded flow of water.

They may be fitted in central-heating systems to enable sections of the system to be isolated. They may also be fitted in the cold-water distribution pipes from the cold-water storage cistern to the hot-water storage cylinder and to the bathroom cold-water taps.

In this situation they make it possible to cut off the hot-water system, or the cold-supply to the bathroom, without putting the remainder of the plumbing system out of action.

2 Ball-valves

The ball-valve could be described in today's advertising jargon as a 'fully automated tap'. Its purpose is to maintain the required level of water in open tanks and cisterns. In the home ball-valves supply water to lavatory-flushing cisterns, the main cold-water storage cistern and – if you have an 'indirect' hot-water system (see Chapter 4) – the small feed and expansion tank connected to the primary and central-heating circuits.

The essential feature of all ball-valves is a float, connected to the valve by a rigid arm of metal or plastic. As the water level in the cistern falls, the float falls with it. The movement of the float arm opens the valve and allows water to flow into the cistern. When the water in the cistern has reached its former level, the float arm closes the valve and stops water flow.

There are a number of different patterns of ball-valve, but the two types most likely to be found in the modern home are the 'Portsmouth' pattern valve and the 'Diaphragm' or 'Garston' pattern valve.

The Portsmouth-pattern valve is the older and more

traditional type. It is always made of either brass or gun-metal and, as can be seen from the illustration, the float arm pushes a washered plug to and fro horizontally in the valve body. When the valve is closed the washer is pressed firmly against the seating of the valve to ensure a watertight joint.

As with taps, washer failure is the commonest fault to which these valves are prone. This is indicated by a steady drip from the overflow pipe of the cistern that projects through the roof or the wall of the house.

To renew the washer of a Portsmouth ball-valve, first cut off the water supply to the valve. In the case of a valve serving the main cold-water storage cistern or a lavatory cistern supplied direct from the main, this is done by turning off the main stop-cock. In the case of the valve of a lavatory cistern supplied from the main cold-water storage cistern, you will have to drain the main cistern and supply pipes as described in Chapter 1.

Some Portsmouth valves have a screw-on cap on the end of the valve body. This must first be removed. Next, with a pair of pliers, pull out the split pin on which the float arm pivots. Remove the float arm and place it on one side. **It is, by the way, a good idea to have a spare split** ◀— **pin handy. If the old one has been in use for some years it may break as you remove it.**

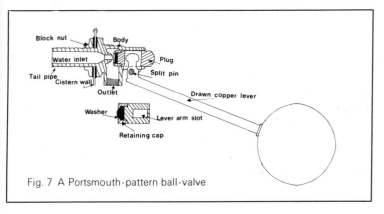

Fig. 7 A Portsmouth-pattern ball-valve

Insert the blade of a screwdriver into the slot in the valve body from which the end of the float arm has been removed, and push the plug out of the valve body. **Have a hand ready to catch it as it comes out.** You won't enjoy fishing for it in fifty gallons of cold water!

The plug is in two parts, although this may not be apparent on inspection. It has a body and a retaining cap that holds the washer in place.

It should be possible to insert a screwdriver blade through the slot in the plug and then to unscrew the retaining cap with a pair of pliers. If this proves difficult – as well it may! – don't struggle too hard or for too long. **Pick out the old washer with the point of a penknife blade and force the new one under the flange of the retaining cap. Make sure that it lies absolutely flat under the flange.**

Before reassembling, clean the plug with fine abrasive paper and – wrapping a piece of fine abrasive paper round a pencil – clean the inside of the valve body in the same way. Apply a light smear of vaseline to the plug before inserting it into the valve body.

Corrosion, and the deposit of hard-water scale, may result in a Portsmouth ball-valve jamming either open or shut. If this happens, dismantle the valve, renew the split pin and clean the plug and interior of the valve body as suggested after washer-changing.

Noisiness is another common failing of Portsmouth ball-valves. Quite apart from the rushing sound of incoming water, there may be the heavy banging of water hammer, or a steady drumming or humming noise as the cistern refills and the valve closes.

Householders are frequently unable to trace the source of this sound. They know only that it occurs after they have drawn off hot water and that it can often be stopped or alleviated by turning on the cold-water tap. This has the effect of reducing water pressure in rising main and diminishing the flow of water through the valve.

These noises are the result of ripple formation on the surface of the water in the cistern as fresh water flows in. The ripples shake the float up and down and to and fro, and this movement is transmitted to the ball-valve and thence to the rising main. Modern copper tubing in particular is liable to act as a sounding board and to magnify the sound out of all recognition.

The more serious phenomenon of water hammer arises from the valve bouncing on its seating, under the influence of the ripples, when it is nearly closed.

In the past it was usual to reduce these noises by fitting a 'silencer tube' into the ball-valve outlet. This tube introduced fresh water into the cistern below the level of the water already in it. However, silencer tubes are nowadays banned by water authorities because of the risk of contamination of the mains by back siphonage.

If you have a noisy ball-valve, try this simple remedy which – while it cannot be guaranteed to prove effective – has helped in a great many cases.

Take a small plastic flowerpot and drill two holes, opposite each other, near to the rim. Tie the ends of a loop of string to each hole. Then slip the loop over the float arm, so that the flowerpot is suspended in the water a few inches below the level of the float.

Here it will have the effect of stabilizing the float and preventing it from bouncing on every ripple.

If this proves effective, replace the string with nylon cord or copper wire that will not rot in constant contact with water.

Another measure that will often reduce vibration and noise is to ensure that the rising main is securely fixed to the roof timbers. This is especially important when replacing an old galvanized steel storage cistern that gives ample support to the rising main with a modern plastic one that does not.

If all else fails, you should consider replacing the

existing Portsmouth ball-valve with a modern diaphragm-valve that can be depended upon to operate more silently.

Portsmouth valves are classified as high-pressure, low-pressure or full-way, depending upon the diameter of the nozzle orifice. HP (high pressure) or LP (low pressure) is usually to be found stamped on the valve body.

High-pressure valves are normally fitted to main cold-water storage cisterns and to flushing cisterns supplied direct from the rising main. Low-pressure valves are used for flushing cisterns supplied from a main cold-water storage cistern. If the main cold-water cistern is situated so that its base is only a few feet above the level of the flushing cistern, it may be necessary to fit a full-way valve with a nozzle orifice of maximum size.

If a high-pressure valve is fitted where a low-pressure valve is required, the cistern will refill much too slowly. If, on the other hand, a low-pressure valve is fitted where there is high water pressure, there will be leakage past the valve despite constant washer-changing.

In some parts of the country mains water pressure can vary greatly during the course of twenty-four hours. Pressure is high at night, when there is little demand for water, and much lower during the day, when demand is heavy. Fit a high-pressure valve in such a situation and the cistern will refill with agonizing slowness during the day. Fit a low-pressure valve and it will leak, causing overflows, during the night.

The answer to this problem is to fit an 'equilibrium' ball-valve. This kind of valve resembles the conventional Portsmouth ball-valve, but it has a channel drilled through the plug permitting water to flow through to a watertight chamber behind it.

An ordinary Portsmouth valve opens under the influence of two forces – the action of the float arm *and* the pressure of water in the main. With an equilibrium valve, pressure is equal on each side of the plug and the valve opens solely under the influence of the float arm.

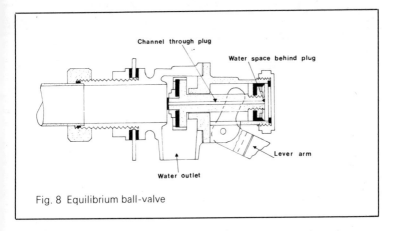

Channel through plug

Water space behind plug

Lever arm

Water outlet

Fig. 8 Equilibrium ball-valve

This makes it possible for the valve to have a much wider orifice without the risk of leakage when water pressure is high and it ensures rapid refilling when pressure is low.

Since the equilibrium valve eliminates the struggle that occurs between the buoyancy of the float and the pressure of water in the main, it also does much to eliminate the 'bounce' that occurs as the valve closes.

Equilibrium valves are, needless to say, more expensive than conventional Portsmouth-pattern valves.

The diaphragm, Garston or BRS ball-valve was developed some twelve years ago at the Government's Building Research Station at Garston in an attempt to produce a silent and efficient valve that would be free from troubles arising from corrosion and hard-water scale (see overleaf).

The diaphragm-valve has a nylon nozzle, resistant to scoring by grit and giving a smooth, even flow of water. The nozzle is closed, not by a moving plug but by a relatively large rubber diaphragm. As water level rises, a small plug presses this diaphragm firmly against the valve nozzle. This plug, which is the valve's only moving part, is protected from the water – and thus from scale and corrosion – by the rubber diaphragm.

When first invented, the diaphragm-valve was fitted

27

with a silencer tube. Since the banning of these tubes by water authorities the manufacturers have developed diaphragm-valves with overhead outlets. These outlets are provided with means of distributing the incoming water in a gentle shower rather than in a noisy stream.

Another recent development is the demountable nozzle which makes it possible to dismantle the valve and convert it from high-pressure to low-pressure use, or vice versa, in a matter of seconds.

A poor flow from a diaphragm ball-valve will usually be found to be due either to the rubber diaphragm jamming against the nozzle or to debris from the mains (or from the main cold-water storage cistern) becoming lodged within the valve.

The remedy, in either circumstance, is to dismantle the valve. Before you attempt this, **cut off the water supply to it.** I stress this point because the poor flow may give the impression that you can dismantle and reassemble it

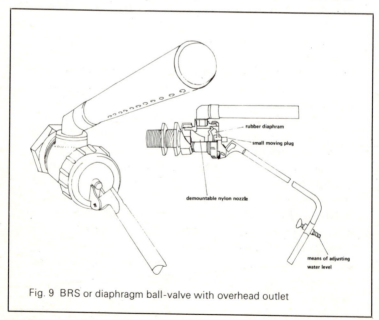

rubber diaphram

small moving plug

demountable nylon nozzle

means of adjusting water level

Fig. 9 BRS or diaphragm ball-valve with overhead outlet

with the water still turned on. You can't! Once the diaphragm is freed or the debris removed a miniature niagara will be released that will make it all but impossible to reassemble.

The valve is dismantled by turning the large ← **knurled retaining nut. It should be possible to do this by hand. The diaphragm can then be picked out with the blade of a penknife or screwdriver and any debris removed.**

Diaphragm-valves always incorporate some kind of screw control that enables the float to be raised or lowered to adjust the level of water in the cistern.

Where a ball-valve is not provided with a control of this kind, the water level in the cistern can be adjusted by bending the float arm.

To do this, first of all unscrew and remove the ← **float. Then take the float arm in both hands and gently but firmly bend the float end upwards to raise the water level or downwards to lower it. Screw the float on again and check that your adjustment has had the required effect.**

A leaking ball-float will make itself apparent by a full-bore flow of water through the overflow or warning pipe.

Never ignore this warning. This 22-mm ($\frac{3}{4}$-in) pipe is not sufficient to cope indefinitely with water flowing into a cistern at mains pressure. Before long the cistern will overflow – possibly with catastrophic results.

The remedy for a leaking ball-float is, of course, to unscrew and remove it and to replace it with a new float. These are readily obtainable from builders' merchants and d-i-y shops, but, unfortunately, this kind of emergency rarely occurs when these premises are open.

To make an emergency repair, tie up the float arm ← **so that water cannot flow in through the valve. Unscrew and remove the leaking ball. With a knife or tin-opener enlarge the area of the leak sufficiently to allow you to drain off the water inside it. Screw it**

back onto the float arm and slip a plastic bag over it, tying the mouth of the bag securely round the float arm.

A running repair of this kind will see you comfortably through the longest Bank Holiday weekend and, when the shops open again, you can buy a replacement float.

3 The Lavatory Suite

A lavatory suite that doesn't function quietly and efficiently can be a serious source of embarrassment both to the householder and to his guests.

An all too common fault that besets the 'direct action' kind of flushing cistern normally provided for modern suites is failure to flush promptly when the flushing lever is operated. Instead of flushing immediately the lever is depressed, two or more sharp jerks are required.

If this happens your first move should be to remove the lid of the cistern and check the water level. There may be a mark on the cistern wall to indicate the correct level. If there is no such mark adjust the ball-float, as suggested in the previous chapter, so that the level of water is about 13 mm ($\frac{1}{2}$ in) below the bottom of the overflow outlet.

Probably though, you'll find that water level is correct. In this case the trouble lies within the siphoning mechanism. Let me explain how this mechanism works.

When the flushing level is operated, a metal plate or disc is raised within the dome of the siphon. This throws water over the inverted U-tube above it into the flush pipe, thus starting the siphonic action. The metal disc has a hole, or holes, in it to permit water to flow through freely once the siphonic action has begun.

When the disc is raised these holes are closed by a valve – usually a thin, plastic diaphragm. This flaps up to allow water to pass upwards through it, but is held down firmly by the weight of water as the disc is raised.

Failure of this diaphragm is the cause of the trouble. It is no longer effectively closing the holes in the metal disc. Thus it is becoming increasingly difficult to throw water over into the flush pipe to start the siphonic action.

To renew this diaphragm you must withdraw the siphoning mechanism from the cistern. Tackle the job this way:

Tie up the float arm of the ball-valve so that water ◀ **cannot flow into the cistern, and flush the cistern to empty it. Unscrew the nut that connects the threaded 'tail' of the siphon to the flush pipe and disconnect the flush pipe.**

Next – and you will need a good wrench for this – unscrew the large nut immediately beneath the base of the cistern. Have a bowl or bucket handy as you do this. The pint or so of water left in the cistern after flushing will escape as you loosen this nut. Once the nut has been removed you will find that the siphoning mechanism can be withdrawn from the cistern and you will be able to renew the defective diaphragm.

It is, of course, wise to buy a new diaphragm before

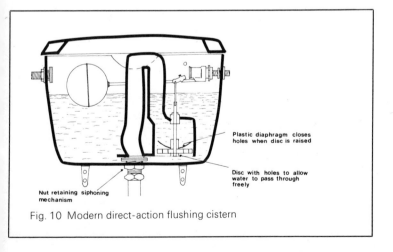

Plastic diaphragm closes
holes when disc is raised

Disc with holes to allow
water to pass through
freely

Nut retaining siphoning
mechanism

Fig. 10 Modern direct-action flushing cistern

beginning the job. **If the shop assistant asks you what size of diaphragm you require, buy the largest. Scissors will easily cut it to size. It should, in fact, be big enough to cover the disc and touch – but not scrape on – the walls of the siphon dome.**

It must be added that although most flushing cisterns are dismantled in the way described there are a few in which the siphon mechanism is secured by bolts inside the cistern. There are other, very modern, ones where the flush pipe is not bolted to the cistern but is thrust into it past a watertight 'O' ring-seal.

Older houses may have, perhaps in an outside lavatory compartment, a high-level lavatory suite flushed by an old-fashioned Burlington or 'pull and let go' cistern.

This kind of cistern is invariably made of cast iron and has a well in its base. In this well stands a heavy, iron bell, to the top of which is connected the chain-operated flushing lever. A stand-pipe, connected to the flush pipe, rises inside this bell and extends, open-ended, to a point an inch or so above full water level.

The flush is set in motion by raising the bell and then

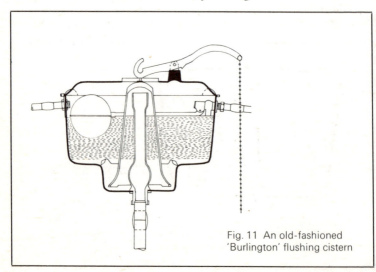

Fig. 11 An old-fashioned 'Burlington' flushing cistern

suddenly releasing it. Its wedge, or conical, shape forces water inside it over the rim of the stand-pipe as it falls heavily into the cistern well. This starts the flushing action. The bell has lugs at its base that permit water to pass under the rim once siphonic action has begun.

Apart from its intrinsic noisiness the principal fault encountered with the Burlington cistern is a tendency for 'continuous siphonage' to occur after it has been in use for some years. When the cistern is flushed it doesn't refill properly. Water continues to flow down the flush pipe into the lavatory pan. This can be stopped only by pulling the chain once again.

The main cause of continuous siphonage is an accumulation of rust, grit and other debris in the base of the cistern well. The lugs at the base of the bell sink into this debris and air cannot pass under the rim to break the siphon after flushing. Contributory causes may be a ball-valve that refills the cistern too rapidly and the slow wearing away of the lugs at the base of the bell.

Cleaning out the base of the bell thoroughly will usually effect a cure. In some cases it may be necessary to reduce the flow through the ball-valve inlet by partially closing a stop-cock in the water supply pipe and in others it may be necessary to build up the lugs on the rim of the bell with an epoxy-resin filler.

Condensation

Both kinds of lavatory flushing cistern may be affected by condensation. Droplets of water appear on the outside surface, giving the impression that the cistern has become porous. Burlington cisterns will rust and, indoors, condensation may be sufficiently severe to drip off the cistern and cause damage to floors and floor coverings.

Condensation occurs when warm, moist air comes into contact with the cold surface of the flushing cistern. The air in bathrooms tends to be both warmer and damper than that in separate lavatory compartments, so it is lavatories situated in bathrooms that are most prone to this trouble.

Improved ventilation gives the most certain answer. In serious cases it may well be worth while fixing an electric extractor-fan in the bathroom window.

Insulating the cistern's surface against the warm, moist air of the room will also help. **Iron Burlington cisterns can be painted externally with two or more coats of insulating anti-condensation paint.**

Plastic cisterns have a built-in insulation that renders them less liable to condensation than ceramic ones. It is sometimes possible to insulate ceramic cisterns internally to cure otherwise intractable cases of condensation.

Empty the cistern and dry it thoroughly. Then line it internally with strips of the kind of expanded polystyrene that is bought in wafer-thin sheets for providing an insulating lining under wallpaper. Use an epoxy-resin adhesive and allow it to set thoroughly before refilling the cistern and bringing it back into use.

Two small final points on condensation: if your lavatory is in the bathroom and is giving trouble in this way, it is worth while to try the effect of running a couple of inches of cold water into the bath before you turn on the hot tap. Try the effect too of refraining from drip-drying washing over the bath. These simple steps may make more radical measures unnecessary.

Cleansing the Lavatory Pan

Failure to cleanse the lavatory pan after flushing is an even more embarrassing fault in a lavatory suite than failure to flush. When the flush is operated two streams of water should flow round the flushing rim with equal force to meet in the centre of the front of the pan. There should be no 'whirlpool effect' as the pan empties.

Here are some points to check on if your lavatory suite isn't properly cleansed with one flush.

Check the level of the water in the flushing cistern. A ful 9-l (2-gal.) flush is essential for thorough cleansing.

Check that the flush pipe connects squarely to the 'flushing horn' of the lavatory pan and that no jointing material or other debris is obstructing the inlet to the pan. At one time it was the practice of plumbers to make this connection with an unhygienic 'rag and putty' joint and putty from such joints was a frequent cause of obstruction. If you have an old-fashioned joint of this kind replace it with a modern rubber cone flush-pipe connector.

Check, with your fingers or with a mirror, that the space under the flushing rim of the pan is free from flakes of rust and other debris that may be obstructing flow.

Check that the pan outlet connects squarely with the socket of the branch drain or soil-pipe.

Finally, with a spirit-level, check that the pan is set dead level. If it is not, **loosen the screws that secure the** ← **pan to the floor and pack underneath with linoleum or similar material to make the pan level.**

Noise

The lavatory suite, like a good Victorian schoolchild, should be seen (when required) but not heard. Unacceptable noise may include the sound of the flushing water descending into the pan, the contents of the pan being discharged into the soil pipe or drain, and the sound of the flushing cistern refilling after use.

Compact building design and the modern practice of confining the soil-pipe taking above-ground drainage within the fabric of the building have tended to increase the audibility of the lavatory compartment.

Measures that may help include ensuring that the joint between the lavatory-pan outlet and the branch soil-pipe is of mastic – not solid – material and raising a floorboard and running an inch or two of dry sand onto the ceiling of the room immediately beneath the lavatory. Ways in which ball-valves may be made to work more silently have already been dealt with and the rush of descending water from the cistern will be less noticeable with a low-level than a high-level suite.

Where silent and efficient operation is an over-riding consideration, the householder should consider replacing the existing lavatory suite with a close-coupled double-trap siphonic model.

With this kind of appliance the flushing cistern and lavatory pan form one unit. There is no flush pipe – not even the short one of a low-level suite – to be seen. Flushing does not depend, as with the common 'wash down' lavatory, upon the weight and momentum of the flush water. Siphonage is employed to empty and cleanse the pan.

The pan outlet has two traps instead of the one trap fitted into a conventional lavatory. A short length of pipe or 'pressure-reducing device' connects the air space between the two traps to the outlet of the flushing cistern.

When the cistern is flushed, water rushing past the end of this pipe aspirates the air out of the air-space between the two traps creating a partial vacuum, at which point

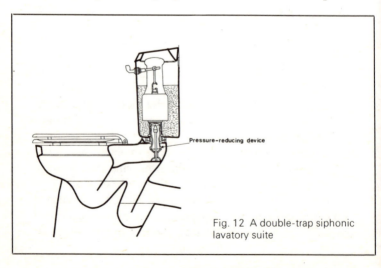

Pressure-reducing device

Fig. 12 A double-trap siphonic lavatory suite

atmospheric pressure – the weight of the atmosphere above the lavatory suite – pushes out the contents of the pan. If the suite is working properly the water level in the pan can be seen to fall before the arrival of the flushing water. This serves largely to wash the sides of the pan and to recharge it.

When considering the purchase of a suite of this kind to replace an existing wash-down suite, it must be remembered that the new suite will project several inches further into the room than the one that it replaces. In some small lavatory compartments or bathrooms this could make the proposed replacement impossible.

Leakage

Leakage from a lavatory-pan outlet when the pan is flushed is a serious fault demanding immediate attention.

It is usually upstairs-lavatory suites that are affected. The joint between the pan outlet and the socket of the branch soil-pipe may be made with putty. With the passage of time the putty dries and shrinks. This, combined with settlement of the floorboards to which the pan is fixed, produces the leakage. Fortunately it is a fault that the householder can easily remedy.

You'll need a tin of non-setting mastic filler and a roll of a waterproofing building tape. Any builders merchant will supply these.

Rake out the existing jointing material with the blade of a screwdriver or some similar appliance and make sure that the space between the lavatory-pan outlet and the soil-pipe socket is dry and clean.

Bind two or three turns of the waterproofing tape ← **round the pan outlet and caulk down hard into the soil-pipe socket. Next, fill in the remaining space between outlet and socket with the non-setting mastic. Finally, bind another couple of turns of waterproofing tape round the completed joint. This can later be painted if required.**

37

A joint of this kind will remain watertight 'for ever' and will readily accommodate slight movement or settlement of the floorboards.

4 Hot-water Supply

There are a number of ways in which a dwelling may be provided with constant hot water on tap. But one version or another of the cylinder storage system is the hot-water system most likely to be encountered in the average suburban home.

The essential feature of a cylinder storage hot-water system is a closed hot-water container, usually – but not necessarily – a copper cylinder, which is supplied with cold water from an open storage cistern at a higher level. The distribution pipe that takes the cold water from the storage cistern to the cylinder should be at least 22 mm ($\frac{3}{4}$ in) in diameter and should connect to the cylinder an inch or so above its base.

A vent pipe, sometimes incorrectly called an 'expansion' pipe, rises from the apex of the cylinder and is taken upwards to terminate open-ended over the cold-water storage cistern. It is from this vent pipe, above the level of the storage cylinder, that the hot-water distribution pipes are taken to the kitchen and bathroom hot taps.

The water in the storage cylinder may be heated solely by means of a thermostatically controlled electric immersion-heater, or solely by means of a solid-fuel, gas or oil-fired boiler. Commonly, a combination of the two methods is used – a boiler heats the water during the winter months; during the summer the boiler is let out and the electric immersion-heater switched on.

Where a boiler is provided, a 28-mm (1-in) flow-pipe is taken from the upper tapping of the boiler to a tapping about one quarter of the distance from the top of the cylinder. A similar return-pipe connects the lower (return) tapping of the cylinder to the return tapping of the boiler. If there is no boiler the flow and return tappings of the cylinders are blanked off.

When the boiler fire is alight there is a constant circulation of lighter, heated water from the boiler and heavier, cooler water from the lower part of the storage cylinder. Hot water, because it is lighter, gallon for gallon, than cold, rises to the top of the cylinder. It is thus always available to be drawn off from the taps. If the immersion heater is switched on, the heated water rises, in the same way, to the upper part of the storage cylinder.

Since the hot-water taps are supplied from a pipe taken from above the cylinder, it is clear that it is impossible to drain the cylinder and boiler from these taps.

A drain-cock must be provided for this purpose. Where there is a boiler, this should be fitted to the lowest part of the return-pipe from cylinder to boiler. Where

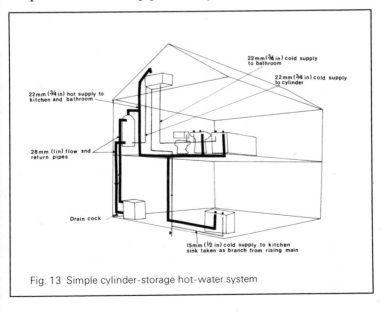

Fig. 13 Simple cylinder-storage hot-water system

there is no boiler and the water is heated solely by means of an electric immersion-heater, the drain-cock is fitted on the cold supply pipe from the cold-water cistern to the hot-water cylinder at a point just before this pipe connects to the cylinder.

It sometimes happens that, to cut the cost of installation, the drain-cock is omitted from a cylinder hot-water system. How, in such a case, can the system be drained in order, for instance, to renew a leaky cylinder or before going on a protracted winter holiday?

Well – it can't be done without creating some mess, but, with a little forethought, this can be cut to a minimum.

Tie up the ball-valve of the cold-water storage cistern and open up all the taps to drain the hot- and cold-water supply pipes. Next, undo the large nut that connects the vent pipe to the dome of the cylinder. Have a bowl and cloth handy as you do this. There is bound to be a pint or so of water spilled.

Next you must siphon out the contents of the storage cylinder. Fill a garden hose with water and arrange for someone to squeeze one end tightly closed over a yard gully, while you thrust the other end deep into the cylinder via the tapping at the apex of the cylinder dome. Release the lower end of the hose and water will flow from the cylinder until it is empty.

Incidentally the hose should not fit too tightly into the tapping at the top of the cylinder. If it does, air will be unable to enter to replace water siphoned off. A partial vacuum will be created and the copper cylinder will collapse like a paper bag under the weight of the atmosphere.

To drain the flow- and return-pipes and the boiler itself there is nothing for it but to undo one of the couplings on the return-pipe. Have a bowl or bucket handy, although unless these pipes are unusually long there will be only a few pints of water to flow away.

Draining a hot-water system in this way is certainly not a job that most householders would wish to tackle more than once. Before refilling, cut the return-pipe from cylinder to boiler at its lowest point. Purchase a drain-cock with compression junctions for connecting to 28-mm (1-in) tubing and fit into the return-pipe at this point. The way in which compression fittings are used is described fully in Chapter 7.

Scale Formation

The formation of hard-water scale or fur on the internal surfaces of boilers and on immersion-heater elements is a serious problem that affects simple cylinder hot-water storage systems in many parts of the country. Areas affected are those with a 'hard' water supply. This means most of southern, eastern and central Britain.

Hardness is due to the presence in the water supply of the dissolved bicarbonates and sulphates of calcium and magnesium. When such water is heated to temperatures in the region of 70°C (160°F) and above, carbon dioxide is driven off and the dissolved bicarbonates of calcium and magnesium are changed into insoluble carbonates which precipitate out as boiler-scale.

Boiler-scale insulates the water in the boiler from the heat of the fire. The householder notices that the water in his cylinder takes longer to warm up – so he piles on more fuel or increases the draught. More scale forms and eventually hissing, bubbling and banging sounds can be heard as overheated water is forced through ever-narrowing channels in the boiler and flow- and return-pipes.

Boiler-scale also insulates the metal of the boiler from the cooling effect of the circulating water. If the mounting evidence of scale formation is ignored, the metal of the boiler will gradually burn away and a leak will develop.

Leaking Boilers

A leaking boiler is among the most catastrophic plumbing emergencies that a householder is likely to have to face. Every householder should know what to do to minimize the mess and to limit the amount of damage.

→ **First – turn off the main stop-cock and open up all the hot and cold taps. Switch off the immersion-heater and put out the boiler fire – if the leak hasn't already done that! Get the garden hose and connect one end to the drain-cock beside the boiler. Take the other end to an outside gully and open up the drain-cock. Make sure that water is running freely through the hose.**

You can now mop up the flood on the kitchen floor in confidence that the system is draining and that very soon the flow from the leaking boiler will cease.

It will – quite obviously – be several days before you can arrange for a new boiler to be fitted. While you are waiting, the hot-water system will inevitably be out of action but there is no reason why the remainder of the domestic plumbing should not be brought back into use.

If there is a gate-valve in the cold-water supply pipe to your hot-water cylinder, you can simply close this valve, open up the main stop-cock and refill the cold-water storage cistern.

The chances are, however, that there won't be such a gate-valve. You will have to find some other way of isolating and cutting off the water supply to the hot-water system.

There will probably be two distribution pipes taken from points near to the base of the cold-water storage cistern in the roof space. One will supply the hot-water storage cylinder and the other the lavatory-flushing cistern and the bathroom cold taps.

→ **Trace the pipe that supplies the hot-water cylinder back to the cistern and push a cork of suitable size into the pipe inlet. Remember that you will have to** withdraw the cork again and that to have it break off in the pipe could be quite disastrous.

To guard against this, screw an ordinary wood screw into the cork before inserting it – leaving 6 mm ($\frac{1}{4}$ in) or so of the screw projecting from the head of the cork. Now you'll be able to be quite confident that the cork can be extracted when the boiler has been replaced and the hot-water system restored to use.

Having cut off the supply to the hot-water cylinder, you can open up the main stop-cock and allow the cold-water storage cistern to refill. Only the hot-water system will be out of action.

Scale Prevention

It will be obvious from the foregoing that everything possible should be done to prevent scale formation in hot-water systems. There are several ways in which this can be done.

Scale doesn't begin to form until water temperature rises above 60°C (140°F). In hard water areas you should set the immersion-heater thermostat to this temperature and, if you have a boiler, you should endeavour to restrict the temperature of the circulating water to this level. This may be possible with a gas- or oil-fired boiler, but you are unlikely to be successful in controlling a solid-fuel boiler to this extent.

The installation of a mains water-softener will reduce all water entering your home to zero hardness and will eliminate absolutely all risk of scale and other hard-water problems. Mains water-softeners are rather expensive, however, and most householders would probably prefer a simpler – and cheaper – solution.

The introduction of certain phosphates of sodium and calcium (sold commercially as 'Micromet') into the hot-water system will prevent scale formation. It is stressed that Micromet does not soften water. Its action is to stabilize the chemicals that cause hardness so that they will remain in solution when the water is heated.

Micromet crystals are placed in a purpose-made basket of plastic mesh. This is then suspended in the water of the cold-water storage cistern a few inches below the level of the ball-valve inlet. Minute quantities of the scale-inhibiting chemical are thus dispensed into the hot-water system and scale formation is prevented. The crystals must be renewed at half-yearly intervals.

Indirect Hot-water Systems

The most permanently effective way of eliminating boiler-scale is by the provision of an indirect hot-water system.

With a system of this kind, the domestic hot water is heated indirectly by a heat exchanger or calorifier within the cylinder. The flow- and return-pipes between boiler and cylinder are connected to this calorifier and this closed system, referred to as the 'primary circuit', is kept supplied with water from a small – say 23-l (5-gal.) – 'feed and expansion tank' in the roof space.

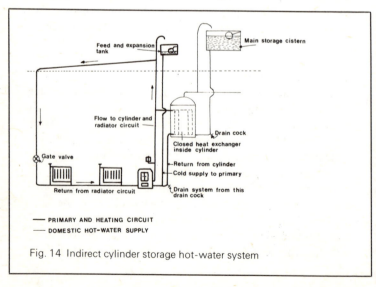

Fig. 14 Indirect cylinder storage hot-water system

An indirect system's immunity to hard-water scale is accounted for by the fact that the water in the primary circuit circulates repeatedly through the boiler and cylinder, and cannot be drawn off except when the whole system is drained. Only the very tiny losses of water due to evaporation are made up from the feed and expansion tank.

When the primary circuit is first filled and the boiler lit, a minute amount of scale is deposited on the boiler's internal surface. This will exhaust the scale-forming chemicals dissolved in the water of the primary circuit. Therefore, as this water continues to circulate, no more scale will form. There may be some scale formation in the outer part of the hot-water storage cylinder, but this will be minimal. The water in this part of the cylinder will become quite hot enough for all domestic purposes but will rarely reach the high temperatures at which boiler-scale is precipitated.

Mention must be made of patent 'self priming' indirect cylinders. These, unlike the conventional indirect system, do not have a separate feed and expansion tank. The heat exchanger consists of a specially designed 'inner cylinder'. When the system is first filled with water from the main cold-water storage cistern, water overflows from this inner cylinder into the primary circuit. An air lock then forms in the inner cylinder to prevent water in the primary circuit returning to mix with the domestic hot water.

The inner cylinder is also designed to provide accommodation for the expansion of the water in the primary circuit when it is heated.

Self-priming cylinders give a less positive separation of primary and domestic hot water than does a conventional indirect system. However they do provide a simple and relatively cheap means of converting an existing direct-cylinder hot-water system to indirect operation.

If central heating is to be combined with hot-water

supply, an indirect hot-water system is essential. Even if hot water only is required, an indirect system is recommended where the local water supply is hard or corrosive.

Air Locks

A common failing of cylinder hot-water systems is the development of air locks. These may occur when the system has been drained and refilled, or they may recur regularly during normal use.

The symptoms are a poor flow, accompanied by bubbling and spluttering from one or more of the hot-water taps.

It usually isn't too difficult to clear an air lock. **Connect one end of a length of garden hose to the nozzle of the cold tap over the kitchen sink and the other end to the hot tap giving trouble. Turn both taps on full.** The cold tap over the kitchen sink is connected direct to the rising main and mains pressure will blow the air bubble out of the system.

If the trouble recurs, you should look for the explanation.

A common cause is having a pipe of too small a diameter feeding the hot-water storage cylinder. This pipe should be at least 22 mm ($\frac{3}{4}$ in) if it is to replace hot water drawn off from the $\frac{3}{4}$-in hot tap over the bath.

If it is smaller than this, water level will fall in the vent pipe when the bath taps are turned fully on. Eventually the level will reach the horizontal run of pipe supplying hot water to the bathroom. Air will enter this length of pipe and an air lock will develop.

Other possible causes of recurring air locks are too small a cold-water storage cistern or a sluggish ball-valve supplying this cistern with water. To check on this, climb up into the roof space and watch the cold-water storage cistern while someone is running off water for a bath.

If the cistern empties before enough water has been

drawn off for the bath, then this will be the cause of the trouble. Either the ball-valve needs to be overhauled or renewed, or a new and larger storage cistern is required.

Instantaneous Water-heaters

Gas multi-point instantaneous water-heaters provide an alternative whole-house hot-water system to the cylinder storage system. These heaters are supplied with water direct from the rising main. This water is heated 'instantaneously' as it passes through a system of small-bore copper tubing within the appliance.

Small, single-point, instantaneous gas or electric water-heaters are also available.

Heaters of this kind give a lower rate of delivery than a cylinder storage hot-water system. They have, however, the advantage that only the water that is actually used is heated. They can therefore be economical where a house is occupied for only a few hours each day.

Small, instantaneous electric water-heaters can be useful in supplying warm water for hand-washing to a lavatory compartment or for providing a shower under circumstances that would otherwise make this im-possible.

The maintenance of appliances of this kind is not, in my opinion, a d-i-y job. If they go wrong, it is wise to call in an expert.

5 Corrosion and Frost

Scale, corrosion and frost might be regarded as the plumbing system's three natural enemies. Scale and its prevention were dealt with in the previous chapter.

The Cold-water Storage Cistern

Is the cold-water storage cistern in your roof space a

traditional one, made of galvanized mild steel? If it is, climb up into the roof space, lift off its lid and flash a torch round its interior. Is it showing signs of corrosion – patches of rust round pipe connections, a light 'dusting' of rust on walls or base, or even warty growths protruding into the cistern from the walls?

The tendency of galvanised-steel cold-water storage cisterns to corrode has increased with the almost universal post-war use of copper supply and distribution pipes. This is the result of an electro-chemical, or 'electrolytic', action that may take place between the copper pipes and the zinc coating of the galvanized steel.

A simple electric cell consists of rods of zinc and copper, connected by a copper wire, immersed in a weak acid referred to as the electrolyte. Bubbles of hydrogen gas form in the electrolyte. Electric current passes from one rod to the other and the zinc slowly dissolves away.

A galvanized-steel cistern to which copper pipes have been connected may reproduce the conditions of an electric cell if the water supply is slightly acid. The zinc coating of the galvanized steel will dissolve away and the steel beneath will be unprotected from corrosion.

If your cistern is already showing signs of corrosion, you can eradicate this and protect the cistern indefinitely by internal painting. Tackle the job this way:

Cut off the water supply to the cistern and drain it from the bathroom taps. You will find that the distribution pipes taken from the cistern are connected two or three inches above its base. This means that you will have to bale – and mop – out the last gallon or so.

➤ **Before proceeding it is best, if not absolutely essential, to disconnect the rising main, the overflow pipe and the distribution pipes from the cistern. This is because it is in the immediate vicinity of the tappings made for these pipes that rust is most likely to develop.**

Dry the walls and base of the cistern thoroughly

and remove every trace of existing rust by means of wire brushing or with abrasives. If you use a wire brush, wear goggles to protect your eyes.

Removing the rust in this way may have left deep pit-marks in the walls of the cistern. It could even have produced holes that penetrate right through the walls in places.

Fill in any such holes or pit-marks with one of the modern epoxy-resin fillers that are available from all d-i-y shops nowadays. Allow to set and rub down to a smooth surface.

Finally, apply two coats of a tasteless and odourless bitumastic paint. Once again your local d-i-y shop will be able to suggest a suitable one. When ordering, stress that it must be non-tainting.

This treatment will give several years' protection against even the most corrosive water supply. It can be repeated if necessary.

A new galvanized-steel cistern can, of course, be protected in the same way at the time of installation. Rub down the walls and base of the new cistern with abrasive paper to provide a key, and apply the paint after the holes have been cut for the pipe connections but before these connections are made.

Anodic Protection

Another way in which a new – or so far uncorroded – cistern can be protected is by means of a sacrificial magnesium anode. Anodic protection depends upon the same electro-chemical principle that produces electrolytic corrosion.

All metals have a fixed and known 'electric potential' and, in the conditions of the electric cell, it is the metal with the higher potential that dissolves away. Zinc has a higher potential than copper – but magnesium has a considerably higher potential than either zinc or copper. A sacrificial anode therefore consists of a lump of

49

magnesium, in electrical contact with the walls of the cistern, submerged in the water.

As electrolytic action takes place the magnesium dissolves away – is 'sacrificed' – and the zinc coating of the cistern walls is protected.

One well-known anode is sold as a lump of magnesium to which is attached a copper wire. A clamp is provided at the other end of the wire.

To fix, scrape the rim of the tank thoroughly down to the bare metal at the point at which the clamp is to be fitted. This will ensure a good electrical contact. Place a batten of wood across the tank and hang the anode over it so that it is suspended well below the surface of the water.

Hot-water Storage Tanks

In the previous chapter I said that the hot-water storage vessel of a 'cylinder storage hot-water system' was

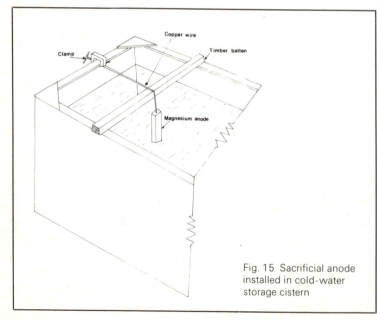

Fig. 15 Sacrificial anode installed in cold-water storage cistern

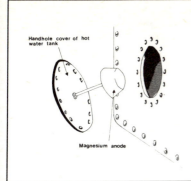

Fig. 16 Installing a sacrificial anode in a hot-water storage tank

usually, but not necessarily, a copper cylinder. Instead of a copper cylinder, many pre-war hot-water systems have a rectangular galvanized-steel tank.

This tank is also liable to attack from corrosion and it cannot be protected by internal painting in the same way as a cold-water storage cistern.

It can, however, be afforded anodic protection. Anodes provided for hot-water tanks normally consist of a lump of magnesium mounted on the end of a metal rod.

To install, tie up the ball-valve of the cold-water ◀— **storage cistern and drain the hot-water system from the taps and from the drain-cock beside the boiler.**

When the system has been completely drained, undo the bolts that secure the hand-hole cover and remove this cover. Drill a 6-mm ($\frac{1}{4}$-in) diameter hole in the middle of the cover and screw on the metal rod so that, when the cover is replaced, the magnesium anode will project into the centre of the tank.

There is one other important point to watch with tanks of this kind. **When a new one is installed, make sure** ◀— **that every trace of metal dust or shaving, resulting from cutting holes for the tappings, is removed before the tank is filled with water. A magnet will help with this, or you can wipe the entire interior surface with a piece of dough.** Any trace of metal dust or shaving left in the tank will unfailingly become a focus of corrosion and will lead to the rapid failure of the tank.

If you have a galvanized-steel hot-water tank of this kind, the pipes connected to it will be either of galvanized mild steel or of lead. Never replace or extend them with copper tubing. To do so will invite electrolytic corrosion. Heavy galvanized-steel tubing is not really suitable for d-i-y use but, if you need to extend such a system, you can safely use stainless-steel tubing. This is almost as easy to handle and install as copper tubing (see Chapter 7), but involves no risk of electrolytic action.

Boiler Corrosion

Rusty red water running from the hot taps – particularly after a large volume of water has been drawn off – *could* be the result of corrosion in the cold-water storage cistern or, if you have one, in a galvanised-steel hot-water tank.

If a check reveals that these are free from rust, then corrosion must be taking place in the boiler. This is a serious matter, as a rusting boiler will eventually leak as surely as will a scaled-up boiler.

The remedy for corrosion in a boiler is the same as that for scale – convert the system to 'indirect' hot-water supply. For corrosion to take place there must be dissolved air present in the water supply. With an indirect system, air dissolved in the water of the primary circuit is driven off when the water is first heated. Although, as will be explained in a moment, there will be *some* dissolved air in the primary circuit from time to time, this will not be sufficient to cause the boiler to corrode.

Corrosion in Central-heating Systems

Small-bore and microbore central-heating systems are customarily based on an indirect cylinder storage hot-water-supply system. It used to be thought that, for the reason given above, such systems would be free of all risk of internal corrosion.

This has not proved to be the case. Some air will always be present in the primary and radiator circuit. It may enter through minute leaks too small to permit water to pass out, and it may dissolve into the surface of the water in the feed and expansion tank.

These tiny amounts of dissolved air are sufficient to permit a kind of electrolytic corrosion to take place between the copper circulating pipes and the pressed-steel radiators of the central-heating system.

The first symptom to be observed by the householder is the fact that his radiators become cool at the top, and need venting more frequently than they did in the past.

If your radiators are needing frequent venting, ⟵ **apply a lighted taper to the vent valve the next time that you open it. If the escaping gas burns with a blue flame, it is hydrogen – a product of corrosion.**

The other product of corrosion is a black iron-oxide sludge, sometimes called 'magnetite'. This obstructs circulation at the bottom of radiators and is drawn through the system to the magnetic field of the circulating pump. Here, its abrasive qualities are a common cause of early pump failure.

Both the hydrogen gas and the magnetite are, of course, produced by the corrosion of the thin walls of pressed-steel radiators. It won't be long before these begin to leak.

Corrosion in central-heating systems can be prevented by the introduction of a reliable chemical-corrosion inhibitor into the feed and expansion tank of the system when it is first installed.

It is possible to introduce an inhibitor into an existing system that is already showing signs of corrosion – but the system must be thoroughly flushed out before this is done.

The boiler must, of course, first of all be turned off or let out. **Then, connect one end of a length of garden** ⟵ **hose to the drain-cock beside the boiler, and take the other end to an outside gully. Open up the drain-**

cock but do not tie up the ball-valve of the feed and expansion tank. Allow water to flow into the system freely to flush it through.

With a padded mallet, tap the surfaces of all radiators – beginning at those furthest from the boiler – to loosen adhering magnetite. Continue to flush the system for at least twenty minutes after the water coming from the hose outlet appears to be clean and clear.

Finally, go up into the roof space and arrange for a helper to stand by the drain-cock ready to turn it off. Hold up the float arm of the ball-valve to prevent water from coming into the feed and expansion tank. When the feed and expansion tank has emptied, signal to your helper to turn off the drain-cock. You can now introduce the corrosion inhibitor into the feed and expansion tank and release the float arm to enable to system to top up.

Light the boiler and switch on the circulating pump for at least an hour to ensure that the inhibitor penetrates to all parts of the system.

It should, incidentally, be noted that when the central-heating system is cold there should be only two or three inches of water in the feed and expansion tank. The water in the primary circuit expands when heated and flows back into this tank to rise *above* the level of the ball-valve float.

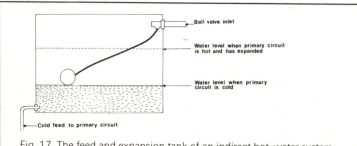

Fig. 17 The feed and expansion tank of an indirect hot-water system

Frost

When water freezes and turns to ice, it expands – increasing its volume by something like 10 per cent. Hence the frozen and burst pipes that accompany any prolonged period of severe frost.

Such periods of severe frost are relatively rare in the British Isles. Because of this we have, in the past, tended to regard frost precautions fairly casually. It has been the need for energy conservation, rather than anxiety about frozen pipes, that has made us take insulation seriously in recent years.

But the threat of frost to the plumbing system is real enough, as we learned during the winter of 1978–9. Even during the much milder winters that we usually experience there are always at least some households caught out by an unexpected spell of sub-arctic weather.

Frost precautions should really be incorporated into the design of the house and its plumbing system. Intelligent design is more effective than the most thorough lagging.

Frost rarely penetrates more than a foot or so beneath the soil of this country, so the 'service pipe', that takes water from the water authority's main to the house, should be at least 0·82 m (2 ft 6 in.) beneath the surface of the soil throughout its length.

Inside the house this pipe is usually referred to as the 'rising main'. A branch will be taken off it to supply the cold tap over the kitchen sink and it will then rise, by the most direct route, to supply the cold-water storage cistern in the roof space.

It used to be considered very important that this pipe should rise against an *internal* wall of the house. Cavity-wall infilling and other means of thermal insulation have made this less important in modern dwellings than it was in the past.

If you live in an older house and the rising main rises against an exposed external wall, lag it thoroughly with fibreglass pipewrap, or with foam plastic or expanded polystyrene pipe-lagging units. **Make sure that the lagging extends behind the pipe.** It is from this direction – from the cold wall – that the danger will come.

In the Roof Space

Modern means of heat conservation may make the plumbing installations in the roof space more vulnerable to frost. Insulating above the bedroom ceilings with fibreglass blanket or vermiculite loose-fill will prevent the roof space receiving heat lost from the rooms below. This may have protected the plumbing in the past.

Pipe-runs in the roof space should be kept well away from the eaves and should be as short as possible. They should be thoroughly lagged and care should be taken to extend the lagging round the bodies of ball-valves and gate-valves. Only the handles of gate-valves should protrude from the lagging.

The cold-water storage cistern should be provided with a dust-proof cover and the walls, but not the base, should be lagged with fibreglass or expanded polystyrene tank-lagging units. **Omitting lagging from the base of the tank and leaving the small area of ceiling immediately beneath the tank uninsulated will mean that some warmth will penetrate upwards from the room below.**

Finally – the cold-water storage cistern's overflow or

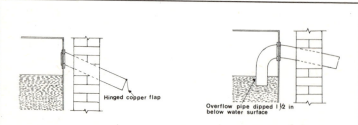

Fig. 18 Two ways of protecting the overflow pipe of a cistern from frost

warning pipe. It used to be the practice to fit a hinged copper flap on the outlet of this pipe to prevent cold draughts from blowing up it into the roof space. **The modern method of protection is to turn the overflow pipe over inside the cistern so that it dips an inch or so below the surface of the water.** The water thus provides a trap that prevents draughts from blowing up the pipe.

Plastic gadgets are available for screwing onto the inlet of an existing overflow pipe to provide this protection.

Winter Holidays

If you are leaving your home empty for more than a few days during a time of the year when frosts are expected, the plumbing system must be protected. Not even the most efficient lagging will afford more than a day or two's protection from severe frost in an unheated, unoccupied house.

If you have a reliable automatic central-heating system the best means of protection is to leave this on under the control of a 'frost-stat' – a thermostat that will switch the heating on when the temperature drops to a prede-termined level.

If you have no central-heating system and a direct cylinder hot-water system (see Chapter 4), the entire system should be drained before you leave home. Turn off the main stop-cock, open up all the taps and drain the hot-water system from the drain-cock beside the boiler or, if the cylinder is heated by electricity only, from the drain-cock at the base of the cylinder's cold-water supply pipe. Leave a large notice by the boiler and the immersion-heater switch: 'SYSTEM DRAINED – DO NOT LIGHT OR SWITCH ON UNTIL REFILLED.'

Incidentally, when refilling the system on your return it is a good idea to connect one end of a length of hose to the cold tap over the kitchen sink and the other end to the drain-cock of the hot-water system.

57

Open the tap and the drain-cock and the system will fill upwards, pushing air in front of the rising water. This will reduce the risk of air locks forming.

The primary circuit of an indirect hot-water system and any central-heating circuit should *not* be drained. To leave a system of this kind empty is to invite corrosion.

Manufacturers of corrosion-inhibitors make anti-freeze solutions that can be added to enclosed primary and radiator circuits to protect them from frost. Automobile anti-freeze should not be used.

Dealing with a Freeze-up

If, despite your precautions, water freezes in your plumbing system, deal with the freeze-up promptly. At first there will be a small ice-plug, easily dealt with, in the pipe. Neglected, this will quickly spread.

A rough indication of the position of the ice-plug can be obtained by noting which parts of the plumbing system have ceased to function. If, for instance, water is flowing into the cold-water storage cistern in the roof space but no cold water is reaching the bathroom, then the ice-plug must be in the cold-water distribution pipe between the storage cistern and the bathroom.

Strip the lagging off the affected pipe and apply cloths, soaked in hot water, or a filled hot water bottle. A hair dryer can provide a useful means of directing a stream of warm air to an inaccessible pipe.

Modern copper piping is a good conductor of heat. Warmth will travel along it to clear a blockage some distance from the point of application.

A Burst Pipe

The first indication of a burst pipe may be water dripping through a ceiling. Don't attempt to find the cause before you have taken immediate steps to limit the effects.

Turn off the main stop-cock and open up all the taps in the house. This will drain the system and prevent serious flooding. It might be wise, too, to let out the boiler or switch off the central heating. The chances are that you will be able to switch on again when you have located the burst – but it is best to err on the side of safety. Only after doing this should you go up into the roof space to find the cause of the trouble.

If you have copper pipework joined either by compression or soldered capillary joints (see Chapter 7), the chances are that the leak will be the result of one of these joints pulling apart. It can easily be remade.

Lead pipe may well split when frozen. The orthodox method of repair is to cut out the affected length and to insert a new piece of lead pipe in its place, joining it to the existing pipework with wiped soldered joints.

This is not – for most of us – a d-i-y job!

The householder can however make a d-i-y temporary repair which – while it may not meet with the approval of the local water authority – could well prove to be as effective as the approved method.

To make it, you will need an epoxy-resin repair kit.

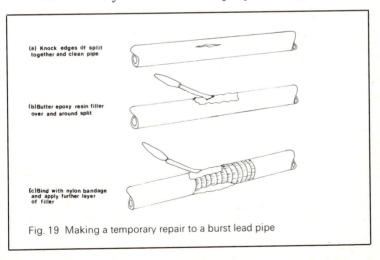

(a) Knock edges of split together and clean pipe

(b)Butter epoxy resin filler over and around split

(c)Bind with nylon bandage and apply further layer of filler

Fig. 19 Making a temporary repair to a burst lead pipe

This is something that every householder should have by him.

→ Knock the edges of the split together with a hammer and clean the pipe with wire wool or abrasive paper. Mix up the epoxy-resin filler according to the makers' instructions and butter round the pipe, over and into the split and for two or three inches on either side of it.

While the filler is still plastic, bind round it a fibreglass or nylon bandage. Over the bandage, through which the filler will now be oozing, butter a further layer of filler.

Within a few hours the filler will have set and you will be able to have water running through the pipe again. When the joint has set thoroughly, it can be rubbed down with sandpaper to give a neat 'wiped joint' appearance.

6 The Drains

The domestic drainage system is out of sight and – until something goes wrong with it – out of mind. A blockage, particularly at a time when you have visitors or when visitors are expected, can be a disaster. Yet most drain blockages can be dealt with promptly and effectively by the householder.

Blocked Waste-pipes

Let's take, first of all, that part of the drainage system that is above the ground.

Any blockage that occurs will almost certainly be in a branch waste-pipe taking the waste from a sink, bath or washbasin to an external gully or to the main soil- and waste-pipe. It will probably be in the trap – the bend in the pipe immediately beneath the appliance that is designed to retain a certain amount of water and thus to prevent drain smells coming back into the room.

Sink wastes, because of the purposes for which a sink is used, are particularly prone to blockages.

The housewife, having completed a pile of washing up, pulls out the sink waste plug – and nothing happens! The sink remains full of greasy washing up water.

A blockage of this kind can usually be cleared quickly and easily by means of a force cup or sink waste plunger. This consists of a hemisphere of rubber or plastic mounted on a wooden handle. Every householder should have one. They can be obtained from any d-i-y shop or household store.

Place the force cup squarely over the sink waste ◄ outlet. With one hand hold a damp cloth firmly against the overflow outlet and, with the other, plunge the handle of the force cup sharply downwards four or five times.

Since water cannot be compressed, the effect of this is to convert the column of water in the upper part of the waste-pipe into a ram which should displace the cause of the blockage. The overflow outlet must be blocked to prevent this force being dissipated up the overflow.

Your first efforts may merely move the obstruction further along the pipe. **If therefore you don't im- ◄ mediately remove the blockage, keep on trying. I**

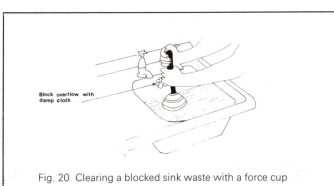

Block overflow with damp cloth

Fig. 20 Clearing a blocked sink waste with a force cup

have known an obstinate obstruction to be success-fully cleared after the waste had been plunged as many as twenty or thirty times.

If you still can't clear the waste pipe, you will have to gain access to the trap. Place a bucket underneath it before you attempt this.

The traditional U-trap has a screw-in cap at or near its base. The entire lower section of the more modern plastic or chromium-plated bottle-trap can be unscrewed and removed. When you gain access to the trap the entire contents of the sink will flow out – hence the importance of the bucket underneath it.

Probe into the trap and waste pipe with a piece of wire. Expanded curtain wire can be very helpful. The chances are that you will find a solid object – a sliver of wood, a piece of bone or something similar – wedged across the pipe.

The last time that I unscrewed and removed the access cap of a trap I managed to break the washer that provides a watertight seal when the cap is screwed home. **I improvised a temporary washer by binding a rubberband tightly round the thread of the cap before replacing it.**

Blockages in bath and basin wastes can nearly always be cleared by plunging. It is not unusual to find that blockages of the outlets of these appliances are caused by

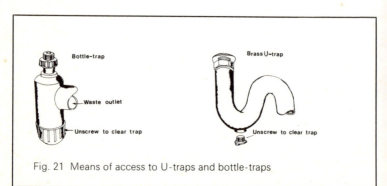

Fig. 21 Means of access to U-traps and bottle-traps

hair caught in the grid. **Try probing through this. A** ←
straightened out paperclip or hairpin can make a
useful tool for this purpose. You may be surprised at
the amount of hair, quite invisible from above, that
is nevertheless suspended from the grid and is
effectively blocking the outlet.

The outlet of a lavatory pan is unlikely to block except
as a result of misuse. To forestall matrimonial discord I
hasten to add that this misuse is not necessarily anyone's
'fault'. Floor cloths, scrubbing brushes and plastic toys –
all potential sources of blockage – have been known to be
'flushed round the bend' in even the best-regulated
households!

A blockage is indicated by the water level in the pan
rising almost to the flushing rim when the cistern is
flushed. It may then, very slowly, subside.

An obstruction of the trap can usually be cleared by
plunging – but a special drain plunger is best for this
purpose. This is a 10-cm (4-in) diameter rubber disc
screwed onto the end of a drain-clearing rod. This is not
the kind of tool that most of us keep about the house, but
an old-fashioned mop – or even a bundle of rags tied, ←
very securely, to a broom handle – can be used in an
emergency.

Alternatively, a length of flexible wire can be passed
round the trap to find and dislodge the obstruction.

However, before attempting to clear an apparently
blocked lavatory-pan outlet in this way it is wise to check
on the underground drains. It is there that the cause of
the trouble may be found.

The Underground Drains

Access to the underground drains is obtained by raising
the cast-iron manhole covers that you will find in your
driveway or in the path at the side of your house.

The cover should have two hand-holds to make it easy
to lift but these could have corroded away if your house is

an old one. **In any event you will find it easier to raise the cover if you first run a screwdriver blade, or the sharp corner of a spade, between the cover and the frame in which it rests to remove the accumulation of dirt that will have found its way there. You can then – if there are no handles – use the spade as a lever to raise the cover.**

Underneath you will find the drain manhole or inspection chamber – a rectangular pit, with walls of brickwork or concrete and a cement-rendered base through which runs the half-channel of the drain. Inspection chambers should be provided at all major changes of direction of the drain and at all points at which a branch drain joins the main drain. You will see branch drains entering the chamber through the wall and sweeping round in a 'three-quarter bend' to join the main drain 'in the direction of flow'.

A drain blockage may be indicated, as already suggested, by a lavatory that fails to flush properly. Other possible indications of a blocked drain are a flooded yard gully or water oozing out from under a manhole cover. With a flooded yard gully, first try raising the grid. It could be that the trouble is due simply to this being clogged with leaves and other debris.

Raising the manhole covers will give you a good idea of the position of the blockage. If one manhole is flooded and the next one is empty, then the blockage must lie somewhere between the two.

You will need a set of drain rods or sweeps rods to clear it. Perhaps you can borrow a set if you don't possess one. **Screw two or three rods together, push one end into the flooded manhole and feel for the half-channel at the bottom. Push the rods along the half-channel into the drain in the direction of the empty manhole.**

Screw on more rods as necessary and keep pushing until the obstruction is encountered and dislodged. To help push the rods along the drain, or to extract

them afterwards, you can twist them in a clockwise direction.

NEVER, however direly tempted, twist in an anti-clockwise direction. If you do the rods will unscrew and some will be left in the drain.

The Intercepting Trap

It used to be the custom to provide the household drains with an 'intercepting trap' to disconnect them from the public sewer. It is many years now since these traps have been installed in the drains of new dwellings but, of course, most British homes *were* built over a quarter of a century ago.

You can check whether or not your house has an intercepting trap by raising the cover of the inspection chamber nearest to the public sewer. This will probably be situated just inside the front gate of your home.

At the end of this chamber the half-channel will appear to discharge into a hole, filled with liquid. Immediately above the 'hole' (actually the inlet to the trap) will be a large stoneware stopper. This stopper seals off the rodding arm which permits the length of drain between the trap and the sewer to be rodded to clear any blockage that might occur.

Near the manhole there may also be a 'fresh-air inlet'. This is a length of drain pipe connected to the manhole and terminating at ground level in a metal box with a grille at the front. A mica flap, hinged at the top, rests against this grille.

The idea of the fresh-air inlet was that air could enter the drain at this point, flush it through and then leave by means of the high-level ventilator or soil-pipe. Air flowing in the other direction – down the drain – would be prevented from leaving by the mica flap.

It rarely worked like that. Fresh-air inlets are ineffective and very subject to accidental damage and vandalism. It is generally better to remove the metal box

65

(if it hasn't already been broken off) and seal the inlet.

The intercepting trap, where it exists, is by far the commonest site of drain blockage. It can usually be cleared by plunging.

For this you will need a drain plunger, although, as for the clearance of a blocked lavatory trap, a mop or a bundle of rags on a broomstick can be used in an emergency.

Assuming that you have some drain rods and a 10-cm (4-in) drain plunger – screw a couple of rods together and screw the rubber plunger disc onto the end of them. Feel for the half-channel at the bottom of the manhole and move the plunger along until you can feel the inlet to the intercepting trap.

Plunge down sharply three or four times. Unless you are very unlucky there will be a gurgle and the water level in the manhole will fall rapidly as the released sewage flows through to the sewer.

After clearing a drain, hose down the sides of the manholes and let the household taps run for half an hour or so to flush the drain through thoroughly.

An unpleasant smell apparent just inside the front garden of a house is usually due to a partial blockage of the intercepting trap. A sudden increase in pressure inside the sewer may cause the stopper to fall out of the rodding arm into the trap to cause a blockage.

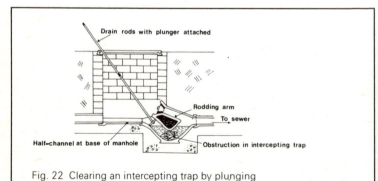

Fig. 22 Clearing an intercepting trap by plunging

Sewage will begin to flood the manhole but will not be noticed because, when it reaches the level of the rodding arm, it can flow down that arm to the sewer. Within the manhole, however, it will become more and more foul until the smell, escaping from the fresh-air inlet or from under the cover, attracts the attention of the householder – or his visitors!

It is always wise to seek the cause of recurring drain blockages. Phone, or write to, the offices of your local borough or district council and ask the Environmental Health Officer if he will call and advise.

Part Two

Plumbing Techniques

7 Handling Copper, Stainless-steel and PVC Tubing

The d-i-y enthusiast who wishes to do more than merely carry out routine maintenance and repair to his plumbing system must master the simple techniques of cutting, joining and bending the tubing that is used in modern hot- and cold-water supply and waste systems.

Copper Tubing

It is the almost universal use of light-gauge copper tubing in post-war plumbing installations that has done most to bring domestic plumbing within the scope of the home handyman.

Half-hard temper copper tubing, as used for domestic water supply and distribution, is available in a variety of sizes but those most likely to be required for d-i-y work are 15 mm, 22 mm and 28 mm. These are the metric equivalents of $\frac{1}{2}$-in, $\frac{3}{4}$-in and 1-in diameter tubing. The discrepancy between, for instance, 15 mm and $\frac{1}{2}$ in. is largely accounted for by the fact that the metric measurement is of the external diameter of the tubing. The old Imperial measurement was of the internal diameter.

To join lengths of copper tubing and to connect this tubing to taps, stop-cocks and ball-valves, the home plumber may use either non-manipulative (Type 'A') compression fittings or soldered capillary joints and fittings. Compression fittings are more expensive but can be used with confidence by the completely inexperienced novice.

A compression coupling, used for joining two lengths of copper tubing, has a joint body with, at each end, a soft copper ring or olive and a cap nut.

To join two lengths of tubing make sure, with a ⟵ file, that the tube ends have been cut squarely and that all 'burr' resulting from cutting has been removed. Unscrew the cap nut from one end of the joint and slip it, followed by the olive, over one of the tube ends. Smear the olive and the tube end liberally with boss white or some similar jointing material. Push the tube end into the body of the joint as far as the tube stop. Now push the olive and the cap nut up to the joint body. Screw on the cap nut hand-tight.

Repeat this process with the other length of tubing and the other end of the coupling.

Finally, use two wrenches or two spanners of appropriate size, to tighten up the two cap nuts. This will compress the soft copper olive against the pipe wall to make a watertight joint.

Most manufacturers insist that with their compression joints a jointing agent is not needed. However, you will

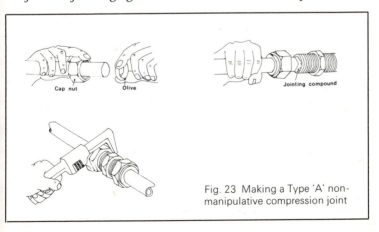

Cap nut   Olive   Jointing compound

Fig. 23 Making a Type 'A' non-manipulative compression joint

find that all professional plumbers use it. It will take up any unevenness in the tube end and ensure a watertight joint at the first attempt.

After a little practice you will probably find that there is no need to remove the cap nut and olive and put them separately over the tube end. You will be able to loosen them and then push the tube end home through them. In the early stages, however, it is wise to make the joint as I have suggested.

Any builder's merchant or d-i-y supplier will be happy to let you browse through his illustrated catalogue of compression joints and fittings and see the wide range that is available. There are 'tee' junctions for connecting branches to existing pipework, reducing fittings for connecting small-diameter pipes to larger ones, a variety of bends, and fittings for connecting copper tubing to the threaded tails of taps and ball-valves.

If – as is quite likely – you need to connect new metric-sized tubing to existing imperial pipework you will find that 15-mm and 28-mm fittings can be used, without adaptation, with $\frac{1}{2}$-in and 1-in diameter tube. Adaptors are necessary for connecting 22-mm tubing to old $\frac{3}{4}$-in pipes but these are readily obtainable.

At some stage you will undoubtedly need to cut a piece of copper tubing to length. This can easily be done with a hacksaw. **Square off the end and remove any 'burr' with a file afterwards. For a big job a tube cutter, incorporating a reamer for the removal of burr, is a worthwhile investment.** It will ensure a square tube end.

Soldered capillary fittings are available in two forms,

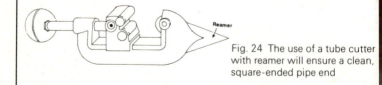

Fig. 24 The use of a tube cutter with reamer will ensure a clean, square-ended pipe end

'integral-ring' and 'end-feed'. Integral-ring fittings incorporate, within a built-in groove, sufficient solder to complete the joint. A separate roll of solder wire is necessary to make an end-feed joint.

Preparation of the pipe ends is as for a compression fitting. **Clean these ends, and the bore of the joint, ◄ with wire wool and smear an approved flux over the tube ends and round the inside of the joint. Thrust the tube ends into the joint as far as the tube stops.**

Then, with the joint firmly supported, apply the flame of a blow torch, first to the tube in the vicinity of the fitting and then to the fitting itself. With an integral-ring fitting, the solder in the ring will melt and will flow, by capillary action, into the narrow space between the wall of the tube and the inside of the joint.

With an end-feed fitting, solder-wire must be held at the mouth of the fitting as it is heated. This too will flow into the narrow space between tube and fitting to complete the joint.

In both cases the joint is complete when a bright ring of solder appears all round the mouth of the fitting. Once made, the joint should be left undisturbed until cool enough to touch.

Where, as with a coupling or a tee junction, two or ◄ more joints are to be made, these should all be made at the same time.

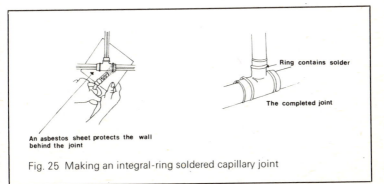

Ring contains solder

The completed joint

An asbestos sheet protects the wall behind the joint

Fig. 25 Making an integral-ring soldered capillary joint

An exact fit is more important with capillary fittings than with compression ones. 15-mm, 22-mm and 28-mm fittings cannot be used with $\frac{1}{2}$-in, $\frac{3}{4}$-in and 1-in Imperial tubes. However, adaptors can be obtained for this purpose.

Although a variety of compression and capillary bends are available it is possible – and much cheaper – to make easy bends in 15-mm and 22-mm copper tubing by hand, using a bending spring of the correct size to prevent the walls of the tube collapsing as you do so.

Grease the spring thoroughly to facilitate removal and insert into the tube to the point at which the bend is to be made. Bend over the knee, overbending at first by a few degrees and then bringing back to the required shape. To withdraw the spring, insert a bar through the loop at the end. Twist to reduce the spring's diameter – and pull. Never attempt to tap or 'dress' the tube until the spring has been withdrawn. If you do you may find that the spring is immoveable.

Stainless-steel Tubing

Stainless-steel tubing is available in the same sizes as copper tubing and can be used for the same purposes. It can be used with copper tubing or with galvanized-steel tanks and tubing without risk of electrolytic action. It is therefore particularly useful for householders who wish to extend or adapt an old galvanized-steel plumbing system.

Although stainless-steel tubing *can* be joined with either compression or capillary joints, I would suggest that the handyman should limit himself to compression joints. For capillary joints a phosphoric acid flux, instead of the usual chloride-based flux, is essential. This can be difficult to obtain and, if you do manage to get hold of some, rather dangerous to handle. It can burn the fingers.

Stainless-steel tubing should be cut with a hacksaw rather than with a tube-cutter and, since stainless

steel is a harder material than copper, the cap nuts of compression joints need to be screwed up rather harder to ensure a watertight joint.

For the same reason I would suggest that spring bending should be attempted only for very easy bends in 15-mm tube. On the whole I think that the novice would be well advised to stick to compression bends when changing direction with tubes of this material.

PVC Tubing

Tubing made of vinyl or PVC (its full name is un-plasticized polyvinyl chloride) has been used increasingly in domestic plumbing in recent years. The home handyman is most likely to find it useful for branch waste-pipes from baths, sinks and washbasins, but it can also be used for main soil- and waste-pipes, roof drainage, underground drains, overflow pipes and cold- (but not hot-) water supply pipes.

It is relatively cheap, light, easily handled and, of course, completely free from risk of corrosion.

The metrication of PVC tubing appears, at the time of writing, to be in a state of confusion. It may be referred to by either its internal or its external metric dimension and many stockists still use the familiar Imperial sizes. When ordering it is probably safest to ask for PVC tubing for sink, bath or basin waste or for overflow pipe as the case may be.

PVC tubing can be joined by means of a kind of compression joint, by solvent-welding or by ring-seal joints. Solvent-welding must always be used for cold-water supply pipes but the householder wishing to fit a short length of waste or overflow pipe, is most likely to find compression fittings useful. Large-scale waste systems frequently use a mixture of compression, solvent-welded and ring-seal joints. The latter must always be used on main soil- and waste-pipes and in underground-drainage systems because they provide accommodation for expansion when warm wastes run through them.

PVC tubing is easily cut with a hacksaw and compression joints resemble those used with copper tubing except that a rubber sealing-ring is incorporated instead of a copper olive. It is never necessary to dismantle a PVC compression joint to fit it. **Loosen the cap nut, slip the pipe end in as far as the tube stop, and tighten up the cap nut again.** It couldn't be simpler.

Solvent-welding PVC tubing is the equivalent of using soldered capillary joints with copper tubes.

Cut the tube end squarely with a hacksaw and remove all swarf and burr with a file. Insert the tube into the socket and mark with a pencil the extent to which it enters. Roughen the pipe end and the interior surfaces of the socket with fine abrasive paper. Do not use steel wool. This will merely polish the surfaces.

Degrease the surfaces that you have roughened, using clean absorbent paper and a cleaning fluid approved by the manufacturers.

With a brush, apply an even coat of solvent cement to the pipe end and the inside of the fitting. Do this with lengthwise strokes, applying a slightly thicker coat to the tube end than to the socket surfaces.

Then immediately insert the tube end into the socket to the pipe stop. Hold in position for about fifteen seconds. You may then remove surplus cement but you should not otherwise disturb the joint for about five minutes, and the pipe should not be brought into use either for cold-water supply or waste drainage for at least twenty-four hours.

Before beginning solvent-welding check whether any printed instructions issued with the fittings vary slightly from the above. One manufacturer, for instance, recommends that the tube end should be chamfered to an angle of 15°. Another suggests that the tube should be twisted as it is inserted into the socket.

Until he has gained some experience the amateur plumber is unlikely to engage in the kind of major plumbing project that involves the ring-seal jointing of large-diameter PVC tubing. The technique is, however, simple enough.

The most difficult part of the job, as far as the novice is concerned, may well be cutting the large-bore pipe end dead square. This, as with any other form of jointing, is essential.

A useful tip is to lay a sheet of newspaper over the ⟵ pipe and bring its edges together underneath. The edges can then be clipped together with a paperclip and will provide a useful 'template' for the saw.

Having cut the end square and cleaned it, mark a line round the pipe 10 mm ($\frac{3}{8}$ in) from the end. Use a rasp or other shaping tool to chamfer the end back to this line.

Next, insert the pipe end into the socket and make a pencil mark to indicate the insertion depth. Withdraw the pipe and make another pencil mark 10 mm ($\frac{3}{8}$ in) nearer to the pipe end than the original one. It is to this second mark that the pipe will ultimately be inserted. This will allow room for thermal expansion and contraction.

Clean the ring-seal recess within the socket and insert the ring. Smear a small amount of petroleum jelly (Vaseline) round the pipe end to lubricate it. Then push firmly home into the socket through the joint ring. Adjust the position of the pipe until the edge of the socket coincides with the pencil mark that you have already made.

Part Three

Some Plumbing Projects

8 Fitting an Outside Tap

No modern householder would dream of using a single power-socket to operate a refrigerator, a washing-machine, a tumble-dryer, a vacuum-cleaner and a food-mixer.

Yet there are many homes in which the cold tap over the kitchen sink, as well as fulfilling its intended purpose for food preparation and washing up, is expected to take a hose for garden water supply and yet another hose for a washing-machine and, perhaps, an automatic dishwasher. The result: inconvenience of course, but also possible damage to an overworked tap. Back pressures from the hose can wash out the gland packing and result in leakage past the spindle and water hammer (see Chapter 1).

Fitting an outside tap – for garden purposes and for washing down the car – is an ideal *first* d-i-y plumbing project. It is a straightforward job within the capacity of any competent handyman. What's more, it can be carried out in two stages, involving minimum interruption of the domestic plumbing services.

An outside tap can be fitted only with the permission of the local water authority. This permission is normally granted readily enough subject to an extra charge on the water rate. It is wise to check on this before you begin.

Assuming that you have the usual 15-mm ($\frac{1}{2}$-in) copper rising main, you will need a 15-mm compression tee, a 15-mm screw-down stop-cock with compression inlet and outlet, a 15-mm wall-plate elbow with a compression inlet and a threaded outlet to take the tail of the tap, two 15-mm compression elbow bends, 1·25–1·5 m (4–5 ft) of 15-mm copper tubing – the exact length will depend upon the position of the new tap – and a $\frac{1}{2}$-in garden bib-tap. The latter should have its handle angled away from the wall so that it can be turned on and off without grazing the fingers.

You will also need a couple of wrenches, a hacksaw, a tin of waterproofing compound, some PTFE thread sealing tape, and means of drilling and plugging the wall to take the wall-plate elbow and of cutting through the wall to take the pipe-run from inside to outside. A suitable electric drill with a hole-cutting attachment is the best tool for this, but the job *can* be done with a cold chisel and hammer.

Decide the level at which you want the tap on the outside wall. Measure up the rising main inside the kitchen to a point about 15 cm (6 in) above that level. Remember that the kitchen floor will almost certainly be higher than the level of the ground outside.

Turn off the main stop-cock. Drain the rising main from the cold tap over the kitchen sink. There *may* be a drain-cock immediately above the main stop-cock. If there is, open it and drain from there too.

Hold the compression tee up to the point that you have marked on the rising main. You will be able to see from *outside* the compression joint how far the pipe ends will project inside. Mark these two points on the rising main – they should be about 18 mm ($\frac{3}{4}$ in) apart – and cut the section between them squarely out of the main.

If there wasn't a drain-cock immediately above the main stop-cock, a pint or two of water will flow out as you do this. Be prepared for it.

Unscrew and remove the cap nuts and olives from the 'run' of the tee junction. Slip first a cap nut and then an olive over the end of the upper length of cut

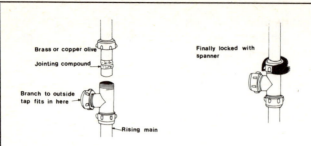

Brass or copper olive

Jointing compound

Branch to outside
tap fits in here

Rising main

Finally locked with
spanner

Fig. 26 Fitting a compression 'tee' into a rising main to provide a supply for an outside tap

pipe. Push them up the pipe and use a spring clothes-peg to stop them from slipping down again.

Slip a cap nut and olive over the lower end of cut pipe in the same way. Once again use a clothes-peg to prevent them from slipping down to the floor.

'Spring' the two ends of the rising main into the body of the compression tee. You will find that there is sufficient 'give' in a length of copper tubing to make this an easy operation.

Make sure that the outlet of the compression tee is pointing parallel to the wall towards the point at which the garden tap is to be fitted.

Apply waterproofing compound to the pipe ends and to the olives. Remove the clothes-pegs, push the olives up to the joint body and tighten up the cap nuts as described in the previous chapter.

Next, take a short – say 15–23-cm (6–9-in) – length of copper tubing and connect this, by the means already described, to the compression outlet of the tee. To the other end of this pipe connect, in the same way, the screw-down stop-cock with compression inlet and outlet.

You will see an arrow engraved on the body of this stop-cock. This indicates the direction of water flow. It is vital that this arrow should point away from the rising main and towards the new outside tap. If it is

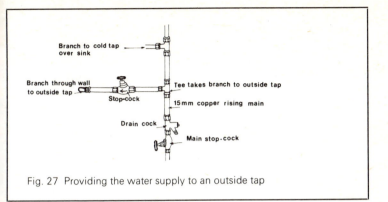

Fig. 27 Providing the water supply to an outside tap

fitted the wrong way round, water pressure will force the jumper down onto the valve seating and no water will be able to pass through it.

Turn this new stop-cock off. Open up the main stop-cock and check for leaks. There shouldn't be any but, if there are, tighten up the compression cap nuts a little more.

The first stage of the job is finished. The domestic water system is back in action again. There is no reason why – if you want to – you shouldn't call it a day and complete the job at some other time.

Next, cut the hole through the wall. It must be large enough to take a 15-mm tube but it shouldn't be *too* large. Don't forget that you will have to make good the wall afterwards with filler.

Measure, and cut off, a length of copper tubing ← **long enough to extend from the outlet of the new stop-cock to the hole in the wall. Cut another piece of copper tubing 32–35 cm (13–14 in) long and connect to the first length with a compression elbow. Push the 32–35-cm (13–14-in) length through the wall and connect the other end of the first length of tube to the compression outlet of the stop-cock.**

Next, go outside the house. Cut the pipe end so that 25 mm (1 in) is projecting from the wall. Connect the other compression elbow to this projecting end so that the outlet of the elbow points downwards towards the position of the outside tap.

Measure the distance from the elbow to the point at which the tap is to be fitted and cut a piece of copper tubing to that length. Connect this piece of tubing to the outlet of the elbow bend and, to the other end, connect the inlet of the wall-plate elbow. The wall must now be drilled and plugged and the wall-plate elbow screwed firmly to it.

PTFE thread sealing tape provides the best means of ensuring watertight screwed joints. It is obtainable in rolls rather like surgical plaster.

Bind a turn of this tape round the threaded tail of your tap and screw into the threaded outlet of the wall-plate elbow. Unless you are very fortunate it is probable that the first time that you do this the tap won't point downwards as it is supposed to. Slip one or more metal washers over the tail of the tap and find out – by trial and error – how many are needed to ensure that the tap is pointed downwards when firmly screwed home.

Your outside tap is fitted! All that remains to be done is to make good with filler the space between the wall and the pipe that passes through it.

One final point – an outside tap could be vulnerable to frost damage. **With the approach of winter it is a good idea to close the new stop-cock that controls the flow**

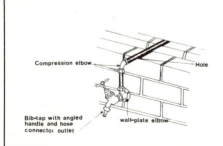

Compression elbow
Hole
Bib-tap with angled handle and hose connector outlet
wall-plate elbow

Fig. 28 The outside tap fitted

to the new tap and to open up the tap itself. This will eliminate any risk of frost damage.

Perhaps it should be stressed that these instructions have been given on the assumption that you have a 15-mm or $\frac{1}{2}$-in copper rising main. Fitting a tee into a lead or galvanized-steel rising main is not, in my opinion, a d-i-y job. However, if you have a rising main of one of these metals there is no reason why you should not get professional help with the actual fitting of the tee and then carry on yourself – in copper if the rising main was of lead and in stainless steel (see previous chapter) if it was of galvanized steel.

9 Plumbing in an Automatic Washing-machine

In many ways plumbing in an automatic washing-machine resembles the provision of an outside tap. Once you have successfully tackled one of these jobs you will have no trouble with the other.

However, an automatic washing-machine needs hot and cold water – so two supply pipes have to be tapped. Provision must also be made for drainage. In many cases the simplest solution will prove to be the best: hook the outlet hose over the rim of the kitchen sink. However, this isn't always possible. Where it can't be done – or isn't convenient – a stand-pipe outlet will have to be provided to get rid of the waste water.

Let's take water supply first. Once again I must assume that you have the usual $\frac{1}{2}$-in (15-mm) copper rising main and a similar $\frac{1}{2}$-in (15-mm) hot-water distribution pipe supplying the hot tap over the kitchen sink.

There is more than one way of tapping the water supply pipes for connection to a washing-machine.

The most common method is exactly as described for tapping the rising main for an outside tap (see previous chapter). For the cold-water connection, turn off the main stop-cock, drain the rising main and insert a 15-mm ($\frac{1}{2}$-in) compression tee at a height convenient for connection to the washing-machine.

The hot-water supply pipe to the kitchen sink must, of course, also be drained before it can be tapped to provide a hot supply to the washing-machine. You will recall from Chapter 1 that, *provided the cold-water supply pipe to the bathroom is taken from a main cold-water storage cistern and not directly from the rising main*, there is no need to drain away all your stored hot water in the storage cylinder.

Tie up the arm of the ball-valve serving the cold-water storage cistern to prevent more water flowing in. Then open up the bathroom *cold taps*. Only when these have ceased to flow should you open up the hot tap over the kitchen sink to complete the drainage of the hot-water distribution pipe.

Cut this pipe and insert a compression tee at the same level as the tee inserted in the rising main for the cold supply to the washing-machine.

Insert lengths of 15-mm ($\frac{1}{2}$-in) copper tubing into

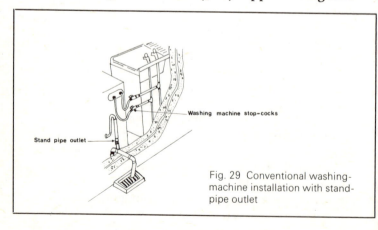

Washing machine stop-cocks

Stand pipe outlet

Fig. 29 Conventional washing-machine installation with stand-pipe outlet

the outlets of these tees. **These should be sufficiently long to take the water supplies to within about 30 cm (one foot) of the machine.** Fit the other ends of these branch supply pipes into the compression inlets of purpose-made washing-machine stop-cocks.

You can get these from any builders merchant or d-i-y supplier. They are attractively finished to enhance, rather than to detract from, the kitchen décor. They have back plates for screwing to the kitchen wall (after drilling and plugging, of course) and outlets designed for connection to a washing-machine hose.

This is, on the face of it, straightforward enough. There *can* be a snag though. The chances are that the hot and cold supply pipes to the kitchen sink run closely parallel to each other down the kitchen wall. This will mean that the branch supply pipe to the washing-machine from the further of these main supply pipes will have to be bent round the other main supply pipe.

There may be enough give in the nearer supply pipe to make it possible to take the branch supply from the further one to the washing-machine without bending it. In any event two easy bends can be made in the branch pipe without too much difficulty using the spring bending technique described in Chapter 7.

However, it is a job that many householders would prefer to avoid and there is a way in which, under most circumstances, it can be avoided. The washing-machine hoses can be connected direct to the main hot and cold supply pipes.

There are at least two ways in which this can be done – one of which even avoids the necessity of cutting the supply pipes. Let's consider the more orthodox method first:

 Kay & Co. Ltd, of Bolton, manufacture the very useful and easily fitted 'Kontite thru-way valve'. These can be fitted into kitchen hot and cold supply pipes and can be connected directly to the hoses of the washing-machine.

The thru-way valve has been designed to make fitting particularly easy. It has Kontite compression joints at each end but one of these joints has not been provided with a tube stop. This means that the valve can be slipped into position without the sometimes difficult job of 'springing' the two cut ends of the copper supply pipe.

You must, of course, cut off the water supply and drain the supply pipes before fitting this valve.

→ **Cut a segment of tube 28 mm (1⅛ in) long out of the supply pipe at the point at which the valve is to be fitted. Square off and clean the tube ends. Slip the cap nuts and olives of the valve's compression joints over each end of the cut pipe, and secure with spring clothes-pegs as suggested in the previous chapter.**

Pull out the upper section of the cut pipe sufficiently to allow you to push it into the joint without the tube stop. Return the upper section of pipe to its former position and lower the valve so that the lower section of cut tube goes into the lower Kontite joint as far as the tube stop. Apply jointing compound to the olives, screw on and tighten up the two cap nuts – and the valve is fitted and ready for use.

Opella plastic hose-connectors can be fitted without the necessity of cutting the supply pipes, although it is, of course, essential that they should be drained before you begin work.

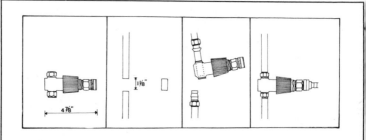

Fig. 30 Fitting a 'Kontite' hose-connector

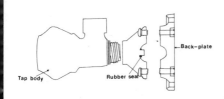

Tap body

Rubber seal

Back-plate

Fig. 31 The plastic
'Opella' hose-connector

Prepare the section of the supply pipe to which the ←
connector is to be fitted by removing any paint and
by rubbing down with fine abrasive cloth.

Carefully drill a 8-mm ($\frac{5}{16}$-in) hole in the centre
of the front of the pipe. You will find that your drill is
inclined to slip, so I suggest that you make a 'pilot
dent' in the tube first, by tapping a nail onto it. Make
sure that the hole is central and do not, on any
account, let your drill bit go through to the back of
the pipe.

Position the back-plate of the hose-connector
behind the hole that you have drilled. Make sure
that the rubber seal is in position round the small
pipe projecting from the front plate. Push this pipe
into the hole and tighten up the screws that connect
the front and back plates.

Drill the wall behind the back plate and secure
with the wall plugs and screws provided with the
hose connector kit.

Bind PTFE thread sealing tape round the threaded
tail of the tap body to ensure a watertight joint and
screw the tap into place. The direction of the outlet
can be adjusted by removing one or more of the
washers provided on the tap tail.

It must be added that although this method of
connection is perfectly satisfactory, the rate of flow of
water into the washing-machine will be rather less than
with either of the other methods described.

If you do not wish to deal with the washing-machine
waste by hooking the outlet hose over the rim of the sink

when required, you can provide a stand-pipe outlet. These are available from manufacturers of PVC tube and fittings. The stand-pipe should be 600 mm (24 in) long and should have an internal diameter of at least 35 mm ($1\frac{1}{3}$ in). This will allow the washing-machine hose to hook into it without giving an air-tight fit.

The stand-pipe must, of course, be securely fixed to the kitchen wall by means of pipe clips.

Where the outlet is taken to a yard gully, a trap at its base is not absolutely essential but a trap must be provided where – as in the kitchen of a first-floor flat, for instance – the waste has to be connected to a main soil- and waste-pipe.

Before making a connection to a soil- and waste-pipe of this kind you should *always* consult the Building Inspector of the local district or borough council.

10 Renewing Sinks, Baths and Washbasins

With all three of these jobs – and with a great many other plumbing projects – removing the old appliance is likely to prove far more difficult than installing the new one.

Let's take the sink first.

Deep, ceramic 'Belfast pattern' sinks with wooden draining-boards retained their popularity for well over a decade after the end of the Second World War. If you move into a house built before or during the years immediately following the war, you are very likely to encounter one in the kitchen.

You will want, of course, to replace it with a modern sink unit having either a stainless-steel or enamelled-steel sink top with integral drainer. Unless price is an overriding consideration, I would suggest stainless steel.

Although enamelled-steel sink tops are attractive in appearance and can be obtained in a variety of colours to tone in with the kitchen décor, they *are* liable to accidental damage. A screwdriver, dropped by an unskilled handyman fixing the curtain rail above the sink, can all too easily chip the enamel. There is really no satisfactory way of repairing damage of this kind.

To avoid an unnecessarily long disruption of the ← **household hot- and cold-water services, it is wise to fit the taps, waste and trap to the new sink before attempting to remove the old one.**

You may decide to fit individual pillar-taps or a mixer. If you decide on a mixer, be sure to specify a 'sink mixer' when ordering.

Bath and basin mixers are, in effect, simply two taps with a common spout. Both hot- and cold-water supplies to the bathroom are normally derived ultimately from the cold-water storage cistern in the roof space. Hot and cold water are therefore under equal pressure. There is no problem about the two streams of water mixing within the appliance.

In the kitchen there is a different situation. The hot-water supply comes, via the hot-water storage cylinder, from the cold-water storage cistern. The cold supply, on the other hand, comes direct from the main.

It is illegal to mix, in any plumbing fitting, water from the main and water from a storage cistern. It is also thoroughly impracticable. The difference in pressure makes proper mixing impossible.

Sink mixers are therefore made with two separate channels inside the appliance – one for hot water and the other for cold. Mixing takes place, in the air, after the water has left the nozzle.

To fit into the sink top, slip flat plastic washers ← **over the tails of the taps or mixer, and insert these tails into the holes provided for them at the back of the sink. Looking underneath the sink you will now**

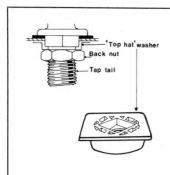

Fig. 32 A 'top hat' or 'spacer' washer must be used when fitting a tap to a modern stainless-steel or enamelled-steel sink

see that, because it is made of thin material, the shanks of the tap tails protrude through. To accommodate these you must slip 'top hat' or 'spacer' washers over the tails before you screw on and tighten up the back nuts.

Modern stainless-steel sinks are provided with combined wastes and overflows, the overflow outlet being connected to the waste by means of a flexible tube. The flange of the waste is usually bedded down into the waste-outlet hole of the sink on a bed of non-setting mastic, although a flat plastic washer is sometimes used.

The trap screws onto the threaded tail of the waste. Traps can be obtained in either brass or plastic. **It may be wise to buy one with a telescopic inlet. This should make it possible to connect the outlet to the existing branch waste-pipe that takes the sink waste to an outside gully.**

Having done as much as you can to the new sink unit you are in a position to progress to the removal of the old ceramic sink.

Remove the wooden draining-board and unscrew the large nut that secures the trap to the waste outlet. Having done this you may be able to lift the sink from the cantilever brackets on which it rests. It is probable, however, that you will first have to chip away, with a cold chisel and hammer, a cement joint connecting the back of the sink to the tiled wall behind it.

You will find that the sink, when freed, is very heavy. Don't attempt to lift it and take it outside without help.

You now have to get rid of the cantilever brackets projecting from the wall. **It will almost certainly prove** ← **easier to cut them flush with the wall with a hacksaw than to attempt to dig them out.**

Unscrew and remove the old trap from the waste-pipe, unless of course, you think that it may be possible to use the old trap with the new sink.

Next – cut off the water supply to the two existing sink taps and drain the supply pipes to them. If you follow the procedures suggested in Chapter 1, only a pint or two of water will drain from them once you have drained the cistern from the bathroom taps. Have a bucket handy to collect this.

The existing taps will almost certainly be bib-taps projecting from the glazed tiles above the sink. The supply pipes to them will be embedded in the wall plaster.

Unscrew and remove the taps. Extricate the supply pipes from the plaster with a cold chisel and hammer, and pull them forward to supply the new taps.

Return to your new sink unit and measure ← **carefully the distance from the floor to the tails of the taps. Measure that distance up the water supply pipes and cut them off squarely at that level. You will probably have to cut another inch or so off them before you have finished, but – in the first instance at any rate – it is better to have them too long than too short.**

Place the new sink unit in position against the kitchen wall. To connect the tap tails to the supply pipes you will need two 15-mm ($\frac{1}{2}$-in) swivel-tap connectors (sometimes called 'cap and lining joints') with a compression inlet. You can get these from any builders merchant.

Insert the 'lining' of the tap connector into the tail of

each tap and screw up the cap nut. The lining is provided with a washer that will ensure a watertight joint without any other jointing material.

Measure the water supply pipes against the compression inlets of the tap connectors. Cut the pipes as may be necessary and connect up the compression joints as suggested in Chapter 7. Restore the water supply and check for leaks.

I have, once again, assumed that the water supply pipes are 15-mm or $\frac{1}{2}$-in copper tubing. If they *should* be of lead, you would be wise to get a plumber to connect the tap connectors to them as wiped soldered joints will be necessary.

All that now remains to be done is to screw the trap onto the waste outlet, connect this to the waste-pipe and, of course, make good the damage that the wall and tiling have undoubtedly suffered.

Basins

Renewing a washbasin resembles, in many respects, the renewal of a sink.

Ceramic basins may be either 'pedestal' or 'wall hung.' Pedestal basins take up rather more room than wall-hung ones but they make it possible for the plumbing to be concealed. It should be noted that modern pedestal basins are always supplied with a concealed bracket or hanger. They should never be supported only by the plumbing and the pedestal.

Ceramic basins normally have a built-in overflow. A slotted waste is fitted to connect this overflow to the main waste outlet. **When fitting this it is important to make sure that the slot in the metal waste coincides with the outlet of the built-in overflow.**

Bottle-traps, rather than simple U-traps, are normally fitted to basin outlets. They take up less room and are neater in appearance. Bottle-traps can be obtained in plastic or in chromium-plated brass.

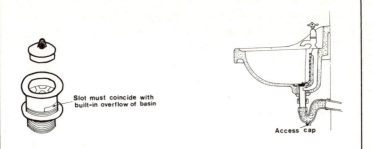

Fig. 33 When installing a ceramic washbasin, the slot in the waste fitting must coincide with the built-in overflow of the basin

Either individual pillar-taps or a basin mixer may be fitted – provided that the bathroom cold supply is taken from a main storage cistern and not from the rising main. Check on this before purchasing a conventional basin mixer.

As ceramic basins are made of relatively thick material, the shanks of the taps will not protrude when the tap tails are inserted into the holes provided for them. **A flat – rather than a 'top hat' – washer should therefore be used between the basin and the back nut of the tap. When screwing up the back nuts make sure that the taps are secure, but do not overtighten. Ceramic basins are easily broken.**

Renewing the taps of a washbasin – as distinct from renewing the basin itself – can be more difficult than appears at first sight.

There is usually no difficulty about unscrewing the cap nuts that secure the water supply pipes to the taps. The back nuts that hold the taps to the basin can, however, be extremely difficult to undo. They are inaccessible and often become virtually welded in position by paint, corrosion and scale.

A cranked basin spanner will make it easier for you to get hold of them and, of course, application of

penetrating oil will help to loosen them.

If, despite the use of oil and the proper tools, you still can't move the back nuts, don't risk damaging the basin. Disconnect the water supply pipes and the waste-pipes. Remove the basin from its brackets and turn it upside down on the floor. You will now be able to get a much better grip on those back nuts and shouldn't have too much difficulty in undoing them.

Vanity units – enamelled pressed-steel basins set in a piece of bathroom or bedroom furniture – are becoming increasingly popular. They are plumbed in exactly the same way as sink units.

Baths

Renewing a bath, even if you do-it-yourself, is a fairly expensive operation. Your first consideration must be the nature of the replacement.

Baths may be made of enamelled cast-iron, enamelled pressed-steel or acrylic plastic. Your existing bath will almost certainly be made of enamelled cast-iron. These baths are strong and hard-wearing – but they are also very expensive, very heavy and consequently difficult to install. The thick material of which they are made also tends to conduct away the heat of the water, making them either expensive or uncomfortable in use.

Pressed-steel baths are cheaper and lighter but they are fairly easily damaged in storage and installation.

My own choice, whether for d-i-y or professional installation, would be a modern acrylic plastic bath. These are light – easily carried upstairs and fitted by one man. They are relatively cheap and, having good qualities of insulation, are economical and comfortable in use. They are tough and hard-wearing. The colour goes right through the material of which they are made and any surface scratches can easily be polished out.

They *are* easily damaged by extreme heat. A lighted cigarette placed, even momentarily, on the bath rim can cause permanent damage. **The d-i-y plumber should** ← **exercise extreme care when using a blow torch in the vicinity of one of these baths.**

Your new acrylic bath will be supplied complete with a wooden or metal frame or cradle, and complete instructions for assembly. This cannot be done until the old bath has been removed.

However, to avoid prolonged disruption of the bathroom plumbing services it is a good idea – as with the installation of a sink or a basin – to fit the taps, and waste before removing the old bath.

Although individual bath taps may be preferred, bath mixers are extremely popular. Many bath mixers incorporate a shower fitting. A flexible metal hose rises from the top of the mixer to terminate in a shower sprinkler intended for attachment to the bathroom wall. A control knob is provided to divert water from the mixer nozzle to the shower sprinkler when required.

Do not invest in one of these before reading the ← **next chapter of this book which sets out the design requirements for successful shower installation.**

Bath taps and mixers are fitted into the holes provided for them in exactly the same way as sink taps and mixers. Acrylic plastic is a thin material and it will usually be necessary to use 'top hat' washers underneath the bath to accommodate the protruding shanks.

Modern baths have combined wastes and overflows, the overflow being connected to the bath trap by means of a flexible hose (see p. 94). Always use plastic traps and waste-pipes with acrylic baths. Rigid metal fittings could damage the plastic material when the bath fills with warm water and thermal movement takes place.

Having fitted up the new bath with its taps and waste outlet you can proceed to the really difficult part of the job: removing the old bath.

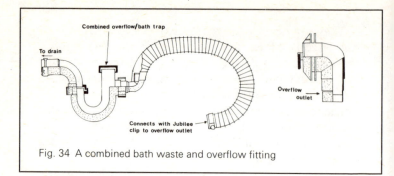

Fig. 34 A combined bath waste and overflow fitting

Strip off the bath panels and have a good look at the water supply, overflow and waste-pipes. They will be in a dark, inaccessible and very confined space between the foot of the bath and the wall.

Cut off the water supply and drain the pipes.

The bath overflow may be connected to the bath's trapped outlet. But in an old installation it is probable that the overflow pipe will be simply taken through the bathroom wall to discharge into the open air outside. **If this is the case, the overflow pipe should be sawn across, flush with the wall, with a hacksaw.**

Although the supply pipes will be inaccessible, it shouldn't be too difficult to undo the cap nuts that connect the water supply pipes to the tails of the taps. Don't attempt to undo the back nuts securing the taps to the bath.

Disconnect the bath trap from the waste-pipe leading outside.

It should now be possible to remove the bath. **Before attempting to do so, lower the bath by means of the screw adjustment on each of the four feet.** This will make it possible to move it without damaging the wall tiling.

The old bath will be very heavy indeed. You will need at least one strong helper to get it out of the bathroom and down the stairs. Don't rush the job. A

Fig. 35 'Copperbend' — either
with or without a tap
connector — can be useful for
making 'out-of-sight' bends
in copper tubing

new bath isn't worth a hernia or a coronary!

Assemble the cradle or frame of the new bath, exactly
in accordance with the manufacturer's instructions, and
place in position. The cradle, if properly assembled, will
prevent the bath from creaking or sagging when filled
with hot water.

It may be that the water supply pipes will be in a
position that will enable you to connect them, by means
of a compression and cap-and-lining connector, to the
tails of the taps without too much adjustment. Don't
forget that the old pipes will be $\frac{3}{4}$-in Imperial size. **You**
will need a conversion fitting (see Chapter 7) to
connect to a 22-mm metric joint.

If the pipes are too short, or are otherwise difficult
to connect to the taps, Ucan 'copperbend' can prove
very useful. This is obtainable in 400-mm and 350-
mm lengths with or without a swivel tap connector
(cap-and-lining joint) at one end. Cut back your
supply pipes to length, fit a length of copperbend
with swivel tap connector to each supply pipe and
you'll find that the copperbend can easily be bent to
reach the tap tails at exactly the right point.

Be sure to make the connections in the right order:
the further tap first, then the overflow and waste-
pipe, and finally the nearer tap.

Do not fit the bath panel until you have restored the
water supply and have checked for leaks.

11 Installing a Shower

Few items of domestic equipment can have been promoted quite so successfully during the past three decades as the shower. Pre-war, a shower was a somewhat exotic exception in a British home. Nowadays no bathroom is complete without one.

Showers save fuel and water, say the manufacturers. You can have four or five satisfactory showers for the same cost as one sit-down bath. Showers are more hygienic and mean less cleaning up for the housewife afterwards. Showers are safer for the elderly and the very young. Showers save space. Conversion of a large old house into small self-contained flats may not leave room for full-sized bathrooms – but an independent shower cabinet can be installed in any space with an area of at least 1 m by 1 m (3 ft by 3 ft).

All this is perfectly true. It isn't, however, always made quite clear that shower equipment that will function perfectly satisfactorily in one home can prove to be an expensive failure in another.

All too often a householder buys and installs an expensive bath/shower mixer set – or one of those basic shower kits with rubber hose-connectors that push onto the bath taps – only to find that the result is a disaster. Perhaps, instead of a refreshing spray, a feeble dribble of warm water drips from the middle of the sprinkler. Perhaps it proves impossible to get a proper mix of water at the right temperature. The shower is icy cold until the cold tap is turned almost completely off – and then becomes suddenly scalding hot!

These troubles result from design faults in the shower installation. For a shower that obtains its hot water from a cylinder storage hot-water system there are certain design requirements that must be met if the shower is to function properly.

The first essential is that the hot- and cold-water supplies to the shower must be under equal pressure. With a cylinder storage hot-water system, pressure on the hot side of the shower will depend upon the height above the shower sprinkler of the cold-water storage cistern. The cold supply must therefore come from the same cold-water storage cistern, not from the rising main.

It is, as I said in the previous chapter, illegal to mix water from the main and water from a storage cistern in any plumbing appliance. If this is attempted with a shower, it will be found quite impossible to get a proper mix of hot and cold water.

The second essential is that pressure must be adequate. The cold-water storage cistern must be sufficiently high above the shower to give an effective spray. Best results will be obtained if the base of the cold-water storage cistern is 1·5 m (5 ft) or more above the level of the shower sprinkler.

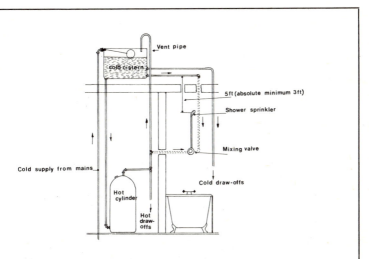

Fig. 36 Design requirements for a conventional shower installation with hot-water supply derived from a cylinder storage hot-water system

If pipe runs are very short with few bends to impede flow, a 'head' of 1 m (3 ft) may prove to be adequate but this is an absolute minimum.

One further – safety – requirement is that the cold supply to the shower should be taken in a separate pipe-line from the cold-water storage cistern. It should not be taken as a branch from some other water-distribution pipe.

If the supply to the shower *is* taken from another supply pipe, there will be a reduction in pressure on the cold side of the shower when cold water is drawn off at some other fitting – at a washbasin or for flushing the lavatory, for instance. This could result in the water from the shower suddenly becoming dangerously hot.

Where the shower is in a bathroom that also contains the only washbasin and lavatory suite, this requirement is less essential. It is unlikely that these will be used at the same time as the shower.

Where pressure is inadequate the best, and cheapest, solution is usually to raise the level of the cold-water storage cistern. This can be done by constructing a substantial platform of timber for it in the roof space.

Having built the platform, cut off the water supply to the cistern and drain it. Cut the rising main and distribution pipes. Raise the cistern onto the platform and insert lengths of copper tubing into the supply and distribution pipes, using the techniques described in Chapter 7.

You may decide that this will provide a good opportunity to get rid of your old galvanized-steel storage cistern and to replace it with a modern plastic one. If so, there are a few points about installation that you should note.

Plastic cold-water storage cisterns – whether made of rigid plastic or flexible polythene – must always rest on a flat level surface. A satisfactory surface can be provided by spiking floorboards across the

wooden frame that you have constructed in the roof space.

The holes for the supply, distribution and over-flow pipes should be cut before the cistern is placed in position. The best method is to use a hole-cutting attachment fitted to a drill or brace. Cut from inside the cistern after positioning a wooden block under the cistern wall to support it.

If you have a very basic toolkit you can mark out the circumferences of the holes with a pencil. Drill a series of small holes round the inside of this mark. Knock out the middle and neaten off with a half-round file.

Pipe connections should always be made squarely to a plastic cistern so as not to stretch or strain the plastic material. Use two washers on each side of the cistern wall to ensure watertight connections. The washer in actual contact with the wall should be of plastic. Do not use boss white or other jointing compound in direct contact with the cistern walls.

Make sure that the rising main is securely fixed to the roof timbers. A plastic cistern does not give the same support to this supply pipe as a galvanized-steel one does. Failure to secure the rising main can result in intolerable noise and vibration as the cistern fills.

There are some circumstances under which it simply isn't possible to raise the cold-water storage cistern. You may live in a flat or ground-floor maisonette. Perhaps you have a 'packaged plumbing system' in which cold-water storage cistern and hot-water cylinder are combined in one unit.

In this case the provision of a shower pump – although it will make a considerable addition to the cost of installation – could be a solution to your problem (see p. 100).

Some homes have no cold-water storage cistern at all. All cold-water services are taken direct from the rising main. The hot-water system consists of a gas multi-point instantaneous water-heater, also fed directly from the main.

Although some gas instantaneous water-heaters can be adapted to supply a shower, the easiest means of providing a shower under these circumstances is likely to be by the installation of an electric instantaneous shower water-heater.

Tremendous strides have been made in the design and production of these appliances in recent years. Installation consists simply of teeing a 15-mm supply pipe from the rising main and connection to a suitable supply of electricity. A control valve regulates the flow of water through the appliance and the temperature of the shower is varied by reducing or increasing the flow.

It must be said that although these appliances can

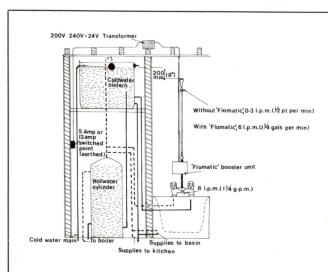

Fig. 37 The 'Flomatic' electric pump unit can be used to boost pressure to a shower where the cold-water storage cistern is too low to permit gravity operation

provide a perfectly adequate shower they have an appreciably lower rate of flow than can be obtained from a conventional shower supplied from a cylinder storage hot-water system.

All showers, other than those supplied by instantaneous heaters, need some kind of mixing valve to mix the hot and cold streams of water to the required temperature.

The most basic kind of mixer is obtained by connecting the streams of water from the two bath taps either in a purpose-made bath/shower mixer, or by means of the simple push-on hose-connectors feeding one of the shower kits that are available from most chemists and d-i-y shops.

Water temperature and flow are adjusted by turning the handles of the bath taps.

This can sometimes be a rather tricky operation. A more positive mix is obtained from a manual shower mixer. These mixers consist of a valve to which hot and cold supplies are connected. Water temperature and, in some instances, volume of flow can be adjusted by turning one knurled control knob. Manual mixing valves are the standard fitting for independent shower cabinets and can also be fitted to showers provided over sit-down baths.

Thermostatic valves are more expensive appliances that are capable of accommodating minor differences of pressure between hot and cold supplies. They will *not* accommodate the very different pressures derived from the main and from a cold-water storage cistern.

They are particularly useful where a number of showers are to be supplied from two main hot and cold supply pipes. In the domestic situation they can perform a useful safety function and they make it possible for the cold supply to the shower to be taken as a branch from a water-distribution pipe serving some other appliance instead of in a separate supply from the storage cistern.

12 The Lavatory Suite: Conversion to Low Level, Renewal of a Pan

Converting a high-level lavatory suite to low-level operation seems to be, on the face of it, a fairly simple d-i-y job. And so it is – provided that the handyman attempting it appreciates *all* the differences between a high-level and a low-level suite. Not everybody does.

Many competent d-i-y enthusiasts have purchased a standard low-level flushing cistern complete with the short, wider-diameter flush pipe – usually sold as a 'flush bend'. They cut off the water supply, disconnect and remove the old high-level cistern and replace it with the new cistern and flush bend – only to find that the new cistern is thrusting forward a foot or so above the level of the lavatory pan. Quite apart from the discomfort of using a lavatory suite with the flushing cistern thrusting at one's shoulders, it will be found to be impossible to raise the lid and the seat completely.

The pan of a conventional low-level lavatory suite is

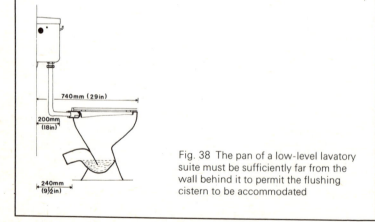

740mm (29in)

200mm (18in)

240mm (9½in)

Fig. 38 The pan of a low-level lavatory suite must be sufficiently far from the wall behind it to permit the flushing cistern to be accommodated

normally situated 5–8 cm (2–3 in) further from the wall behind it than that of a high-level suite. This is to enable the low-level cistern to be accommodated.

Until relatively recently the only way in which a high-level suite could be converted to low-level use was to bring the lavatory pan two or three inches forward into the lavatory compartment. Modern plastic drain connectors and extension pieces have made this a less difficult task than it would have been in the past. It is, however, still a somewhat daunting job and one that, in some cases, involves the replacement of a perfectly satisfactory lavatory pan.

Furthermore, in a small bathroom, bringing the pan forward – even two or three inches – can mean that it is unacceptably close to the bathroom washbasin. The washbasin too must be moved – possibly to a position that will make it impossible to open the bathroom door!

Fortunately, during recent years, slim-line flushing cisterns, or 'flush panels', have been developed that simplify high- to low-level conversion. In the great majority of cases the same lavatory pan can be used, in its existing position.

With a flush panel, the flushing inlet of the lavatory pan can be as little as 13·3 cm (5$\frac{1}{4}$ in) from the wall behind it compared with the 20–23 cm (8–9 in) required by a conventional low-level cistern. In order to accommodate the full two gallons of water required for an adequate flush, these slim-line cisterns are rather wider than conventional ones. Before purchase, make sure that there is sufficient *width* of unobstructed wall behind the pan.

If the high-level suite that you are converting was installed prior to the Second World War it is possible that the flush pipe will be connected to the lavatory pan with a 'rag and putty' joint. These are thoroughly unhygienic and should be replaced with a new rubber cone joint. This consists of a rubber cone, open at both ends. **The larger end is pushed over the flushing inlet to the** lavatory pan, and the end of the flush bend is pushed into the smaller end so that it connects squarely with the inlet. If necessary both ends of the cone can be bound round with copper wire to ensure watertight connections.

You may feel, of course, that the old lavatory pan *should* be replaced. A pan with a crazed or cracked surface is a potential health hazard and should be renewed without delay, whether or not conversion from high level to low level is planned.

How difficult it is to remove the old pan and to replace it with the new one will depend to a large extent upon whether the suite is installed upstairs or downstairs. The pan of an upstairs lavatory suite will probably be connected to the branch soil-pipe with a putty or mastic joint. The pan itself will be fixed to the wooden floor of the bathroom or lavatory compartment with brass screws.

→ **Disconnect the flush pipe from the pan. Undo the brass screws that secure it to the floor and pull forward. If you move the pan from side to side as you do so, its outlet should come out of the socket of the branch soil-pipe without too much difficulty.**

Clean all existing jointing material out of the soil-pipe socket and place the new pan in position, making sure that its outlet goes squarely into the centre of the socket.

You have a choice of methods for making the joint between the outlet of the new pan and the branch soil-pipe. You can, if you wish, use the method suggested in Chapter 3 for dealing with a leaky joint between pan and soil-pipe.

Bind two or three turns of waterproofing building tape round the pan outlet and caulk down hard into the socket. Fill in the space between outlet and socket with a non-setting mastic and complete the joint with another two or three turns of building tape.

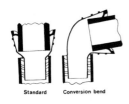

Standard   Conversion bend

Fig. 39 'Multikwik' plastic drain-connectors provide a simple means of connecting a lavatory pan to a branch drain or soil pipe

You may, on the other hand, feel that the use of a modern plastic push-on drain-connector – such as the 'Multikwik' connector – is less messy and gives a neater completed finish. You can get these from all builders merchants. There is, in fact, a range of drain-connectors and extension pieces that will meet virtually any situation in which you may find yourself.

Slip lead washers over the fixing screws before you ◄ insert them into the holes at the base of the lavatory pan to secure it to the floor. These will protect the ceramic surface when you drive the screws home.

It is vital that the pan should be set dead level. Check this with a spirit-level and, if necessary, pack slivers of wood or pieces of linoleum under the lower side to bring it level.

Use a rubber cone connector to connect the flush pipe to the flushing inlet of the new pan.

Renewing a ground-floor lavatory pan can be a good deal more difficult. The pan is likely to be cemented to a solid floor. The S-trap outlet will probably be connected to the branch drain with a cement joint.

Tackle its removal this way. Disconnect the flush ◄ pipe from the pan. With a hammer, deliberately break the lavatory pan outlet behind the trap.

Fig. 40 To remove and replace the pan of a ground-floor lavatory suite it will usually be necessary to break the pan outlet just behind the trap

To renew pan, break outlet at this point

Loosen the front part of the pan from the floor by driving a cold chisel between its base and the floor. Pull the front part of the pan forward out of the way.

You will now be left with the jagged end of the pan outlet projecting from the socket of the branch drain. **Stuff a wad of cloth or newspaper into the drain to retain fragments of cement and ceramic material.**

Attack the protruding end of the pan outlet with a hammer and a small cold chisel. Work carefully, keeping the blade of the chisel pointing towards the centre of the pipe. Break down the outlet right to its end at one point and you will then find that the rest of it will come out fairly easily.

Attack the jointing material in the same way. Make every endeavour not to break the drain socket, although, if you do accidentally do so, the situation is not entirely hopeless. The push-on drain-connectors, to which I have already referred, can be used in a drain with or without a socket.

Having cleared the socket, carefully remove the wad of cloth or paper, doing your best to make sure that no debris falls into the drain.

The new lavatory pan should not be cemented to the

floor. It is better to secure it with fixing screws only. **Chisel all existing cement mixture from the floor to** ← **obtain a smooth surface.**

Now, place the new pan in position with its outlet projecting squarely into the socket of the drain. It must be positioned centrally with an equal space between the outside of the pan outlet and the inside of the socket all round.

If you have managed to break the socket in your efforts to clean it out, you will have to use a push-on plastic drain-connector.

If the drain socket is intact, you can use a drain connector, a mastic joint, or – if you prefer – you can make another cement joint. However the joint is made, the floor must first be prepared for the fixing screws. If you are making a cement joint, the pan must be finally fixed to the floor first. With either of the other joints the final securing of the pan to the floor should be completed after the joint is made.

With the pan in position, mark through the screw ← holes to the floor with the refill of a ballpoint pen. I stress a refill because the holes will probably be too small to allow the actual pen to pass through them.

Remove the pan; drill and plug the floor at the four points marked. It is now important that the pan should be placed back in exactly the same position as before. Marking the position of the pan on the floor before you remove it, and probing through the screw holes to ensure that the plugs are directly beneath them, will ensure this.

Another method is to straighten out four paper-clips and push them into the centre of the plugs. Now lower the pan so that the paperclips protrude through the screw holes at its base and you must have it in the right position.

Slip lead washers over the screws, insert through the holes into the plugs and screw firmly home. As you do so, check the level of the pan with a spirit-level and pack underneath with linoleum as may be necessary.

→ To make a cement joint, dampen some newspaper and caulk down hard into the space between the pan outlet and the drain socket. This will prevent jointing material squeezing through the joint into the drain – a frequent cause of drain blockages.

Make up a mixture of two parts of cement to one part of clean, sharp sand. Fill the space between pan outlet and drain socket with this mixture, making sure that there are no 'voids'. Finish off to a neat, angled fillet at the top of the socket.

A cement joint of this kind must be left absolutely undisturbed for twenty-four hours. Do not disturb the pan in any way – even to fix the lid and seat, or to connect the flush pipe – until after the cement has set thoroughly.

Index

Acrylic baths 92–5
Air locks 46–7, 57–8
Antifreeze solutions 58
Automatic washing-machines
    81–6

Ball-valves 13, 22–30
Basin mixers 87, 91
Basin spanner 91
Basin waste 90–1
Bath mixers 87, 93
Baths 92–5
Bath wastes 93
Bib-taps 11–12, 77, 80
Blocked drains 63–7
Blocked lavatory pan 63
Blocked waste pipes 60–3
Boiler corrosion 52
BRS ball-valve *see* diaphragm
    ball-valve
Burst pipes 58–60

Cap and lining joints 89, 95
Capillary joints 59, 68, 70–2
Central heating 45–6, 52–4, 57–8
Cold-water storage cistern
    47–50, 56–7, 97–100
Compression joints 59, 68–70,
    72–3, 76–80, 82, 89–90, 95
Condensation 33–4
Copperbend 95
Copper tubing 68–72
Corrosion 47–54, 58
Corrosion inhibitors 53–4

Diaphragm ball-valves 27–9
Drain cocks 39–40, 57, 77

Draining a hot-water system
    39–41, 57, 82
Drain inspection chambers 63–7
Drain smells 66–7

Easyclean covers 14
Electric immersion heaters 38,
    42–3
Electrolytic action 48, 49–54
Equilibrium ball-valves 26–7

Garston ball-valves *see*
    diaphragm ball-valves
Gate valves 21–2, 42, 56
Glands (tap and stop-cock) 12,
    19–20

Hard-water scale 24, 41–5
Hot-water storage cylinders
    13–14, 38–47
Hot-water supply 38–47

Indirect hot-water systems
    44–7, 52
Intercepting traps 65–7
Instantaneous shower heaters
    100–1
Instantaneous water heaters 47

Jammed stop-cock 21
Jumpers (tap) 11, 15

Kontite thru-way valve 83–4

Lagging 55–6
Lavatory pan, renewal 104–8
Lavatory suites 30–8, 102–8

Leakage from lavatory 37–8
Leakage past tap or stop-cock
    spindle 19–21
Leaking ball-float 29–30
Leaking boiler 41–3

Manhole covers 63–5
Manual mixing valves 101
Metrication 68, 70–2, 73, 95
Micromet scale inhibitor 43–4
Mini-stop-cocks 21
Multikwik drain connectors 105

Noisy plumbing 24–5, 33, 35–7

Opella hose connectors 84–5
Outside taps 79–81
Overflow pipes 56–7

Portsmouth ball-valves 22–6
Primary circuit 44–5
PTFE tape 80, 85
PVC tubing 73–5

Re-washering:
    ball-valves 23–4
    taps 12–18
Ringseal joints 73, 75
Rising main 55, 76, 97

Sacrificial anodes 49–51
Self-priming indirect cylinders
    45
Service pipe 55
Showers 96–101

Shower pumps 99–100
Shower mixers 101
Shrouded heads (taps) 15–16
Silencer tubes 25, 28
Sinks 86–90
Sink mixers 87
Siphonic lavatory suites 36
Solvent-welding 73–4
Stabilizers 25
Stainless-steel tubing 72–3
Stop-cocks 12–13, 20–2, 76,
    78–9, 80
Stuffing box *see* gland
Supataps 16–18

Taps 11–20, 76–7, 80–1, 87–9,
    91–3, 95
Thawing out pipes 58
Thermostatic mixing valves 101
Top-hat washers 88, 93
Traps 60–3, 88
Tube bending 72
Tube cutting 70–5

Valve seatings 18

Wash basins 90–2
Washers:
    ball-valve 23–4
    tap 12–18
Washing-machine stop-cocks 83
Washing-machine waste 82,
    85–6
Water hammer 19, 25
Water softening 43